The Economic Management of Physical Assets

The Economic Management of Physical Assets

by

N W Hodges RD**, BSc, CEng, MIEE, MIMechE

Mechanical Engineering Publications Limited
London and Bury St Edmunds, UK

First published 1996

ISBN 0 85298 958 X

A CIP catalogue record for this book is available from the British Library.

Printed in Great Britain by Antony Rowe Ltd, Chippenham, Wiltshire

Contents

performance requirements – Financial rules – Source of finance – Discount factors – Mutual influences of projects – External influences

Project life – Asset life – Component lives – Life limited components – Replacements – Economic choices – Life phases and principal features – Viewpoint of user organization

Comprehensive scope of asset – Inclusion of essential buildings, storages, handling facilities, technical manuals, drawings, training aids – Totality to be considered in assessments

Costs/incomes spread over project life-time – Value affected by timings and inflation of currencies – Correction of actual cash flows to 'real values' and common basis of values – Present worths – Discount rates – DCF – Risks – Rates of return – Factors influencing discount rates

Varying scope of technique – Comparisons with full economic assessments – International standards work – Consequential costs of failures – Problems of application

Effects from users adopting economic management – Supplier's reputation – Past performance and attitude – Advertising efforts on customers' behalf – Value and functional analyses in design – Production techniques – Importance of supplier/user communications – Market research

Sponsor in user organization – Need for product – Likely demand – Functional requirement of asset – Capacity and performance – Development of concept – Competition between prospective projects – Formal submissions – Common formats and sensitivity analyses – Authorization – Project manager appointment – Specification and

Coordination with maintenance – Fault location – Exercising emergency procedures – Training for hazards – Effects on environment – Performance monitoring – Site emergency plans

The objective – Balancing maintenance cost and performance achieved – Economic maintenance – Consequential costs – Minimizing need for shut down maintenance – Corrective and preventive maintenance – Calendar or condition based maintenance. On-load preliminary work – Repair by replacement – RCM – Instruction cards – Feedback and reports – History files – Investigation of major failures – Recommended spares lists – Strategic and routine spares – Life limited components – Lubrication

Further use possibilities – Reuse or sale – Separation of scrap materials – Project life extension – Safety aspects – Specialist sub-contractors – Contamination – Hazardous materials – Demolition of buildings

About the Author

After graduating from Bristol University (BSc (Hons) in Engineering), Bill Hodges was called up for National Service in the Royal Navy and served as an Engineer Officer. On demobilization he joined the Electricity Supply Industry. Initially he was employed on operation and maintenance in conventional generating stations, but on the commencement of the nuclear programme was appointed to a post at Berkeley Nuclear Power Station, the first of the commercial stations to generate. In 1962 he took up a post in the Operations Department of the CEGB Headquarters, where he served until retiring in 1988. In this department he was initially engaged in dealing with operational and maintenance problems arising at the nuclear power stations, and liaison with the design departments. Later this work was extended to cover problems arising in all types of station industry-wide. These included such problems as pulverized fuel explosions, statutory requirements for the examination of plant, and the policy of using strategic National Spares for generating and transmission plant. Arising from this work, he became the Electricity Supply Industry's representative on the British Standards Institution's Terotechnology Committee. He became Chairman of this committee in 1985 and, at the request of the committee, remained in the position after his retirement from the industry.

Bill is a member of both the Institution of Electrical Engineers and the Institution of Mechanical Engineers, and is also a member of the British Nuclear Energy Society. He has written, or co-written, several papers published by these institutions dealing with a wide range of topics. He is also the author of *Strategic Spares and the Economics of Operations* (Mechanical Engineering Publications Limited, 1994).

Following his National Service, he served for over 30 years in the Royal Naval Reserves, retiring in the rank of Commander RNR in 1987, having been awarded the Royal Naval Reserve Decoration and two clasps.

Chapter 1

Introduction

General

Any organization or individual undertaking a new project, will wish to plan it and carry it out in a manner that is most economically advantageous to themselves. A project will invariably have a product which may be material in nature or may be a service or even something more esoteric such as a social service or simply just satisfaction. In all cases the product will be deemed to have a value to the intended recipients: these may be 'customers' or may be the supplying organization or individuals, themselves.

Organizations, projects and products can obviously vary enormously but many of the principles of management, including those to be discussed in this book, will apply in almost all cases. As a result it will be necessary throughout to express ideas using generalized terms such as 'product' or 'project'. These will be defined and explained in Appendix I.

In order to carry out a project and produce the product, the organization or individual concerned will invariably need some tools or equipment with which to do so. The total collection of such items needed to carry out the project and to convert the input, 'feed materials' into the desired 'product' will, for this book, be termed 'the asset'. This asset could be plant or equipment and will include any necessary buildings or structures needed to house or support it.

When the organization concerned is a manufacturer, the elements involved in the project are fairly self evident. The 'project' is to 'make the nominated product and to sell it to the customers'. The 'asset' comprises the fabricating machinery, the factory in which this is housed and all the other items needed to carry out the project. This might include stores, transport

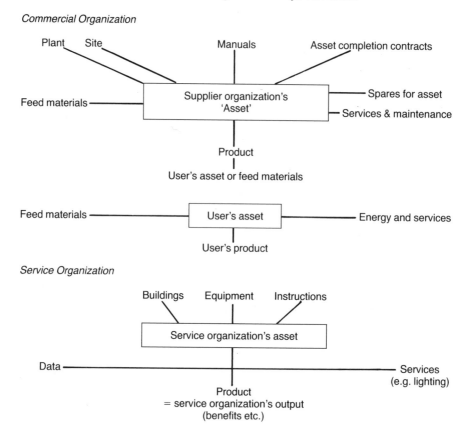

Fig. 1.1   Examples of organizations' inter-relations

and manuals for the operation and maintenance of all the parts of the asset. In this case the 'feed materials' are the raw materials which are worked upon to produce the 'product'. In different types of organization, such as a social services office, whilst the corresponding elements may not be so apparent, equivalents can usually be found. Taking such an organization as an example, the 'asset' would be the office buildings, their services (heating and lighting, etc.) and any equipment, (computers, files, etc.). In this case the 'product' is the service given to 'customers', the beneficiaries of the service and the 'feed materials' would be principally the data input. These two examples are shown in the flow charts in Fig. 1.1.

Economic optimization

In order for the organization to operate to its best economic advantage, it is essential that it manages the asset it has acquired for any project in a manner that is as close as is practicable to the economic optimum that is possible. What is the best economic advantage will depend upon the organization and the project involved. For a commercial organization, this will normally mean in a way that achieves the maximum profit for the organization. On the other hand, for a public service it will mean achieving the desired output or service at the minimum cost to the tax-payers funding it. As this book is devoted to discussing the management of the assets, in either case the organization is the 'user' of the 'asset' and will therefore be termed the 'user organization' to differentiate it from the organization manufacturing, constructing or supplying the asset which will be identified by the term 'supplier organization'. At the same time reference to 'managing' the asset will include its selection, acquisition, its use or operation, its maintenance, and finally its disposal and replacement, should this step be required. In other words, its management will extend throughout its entire 'life cycle' from the time that the project is first conceived until the asset is finally disposed of. The main difference between what we shall call 'commercial' and 'service' organizations are shown in Table 1.1.

Seeking the economic optimum results from a project implies that the assessment of the project, as a whole, or comparisons of alternative courses of action have to be made in economic, essentially financial terms. In addition such assessments must take into consideration all costs and/or incomes from the project throughout the entire life cycle of the asset. When the project is first conceived and during the early phase whilst the asset is being sought and acquired, most assessments will be aimed at identifying the total costs and the total likely income from the project to establish its overall viability. As alternative courses of action become apparent, the choice between these should be made on the basis of whichever shows the better economic effect over the residual life of the project. That is to say whichever produces the better profits or lower costs, for the project as a whole and dependent upon the type of user organization involved, as outlined above.

In many cases the asset will be purchased by and then subsequently owned by the user organization. However, alternative methods of acquiring the asset or parts thereof are sometimes possible. Where such alternatives are available, the choice of acquisition method should take into consideration the economic advantage of one method over the others. At the same time the practicality of the individual methods must be con-

Table 1.1   Typical characteristics of the two types of user organization

'Commercial' (income generating)	'Service'
Objective: To make profit on the project	Objective: Not to make profit but to achieve declared output performance at minimum cost
Financed by money from contributors, to be repaid with interest	Often financed out of public money raised by taxation (e.g. defence, household waste disposal)
Income directly related to the output of the product (e.g. washing powder, dishwashers)	Income, if any, usually has an element of 'standing charges' and thus not directly related to product output e.g. gas, telephones. Others are non income generating services e.g. police, benefit agencies
Asset design based on economic balance between performance and costs	Asset design directly constrained by need to meet performance requirements
Go ahead for project dependent on economic viability	Go ahead with project mainly dependent upon an agreed need

sidered. For example, if adequate financing of an outright purchase of the asset cannot be secured, hiring or leasing may be the only practical option, even if this might be less advantageous financially. It will be clear that the method(s) of financing the project will be a factor in the economic assessments. There will also be considerations of tax and grants or other financial assistance as well as the requirements for writing down the asset. This will be discussed in more detail later.

Historical background

Following the Second World War, as after any period of enormous economic or social upheaval, there was a need, in many industries, including construction, for a major production effort to replace assets lost in the conflict. There was also an urgency to make up for the production lost through resources being diverted to the war effort. At the same time, advances in technology, often stemming from the research and developments carried out in this period, led to designs for plant and buildings becoming more technically advanced than pre-war. These two factors led to requirements for maintenance of major assets which were both increased in extent and more sophisticated in nature. In many cases this resulted in enhanced maintenance costs concurrent with a lower availability of the assets.

Blame for this situation was often unfairly placed upon the maintenance function. Little account was taken of the factors mentioned above or of the inescapable facts that the 'quality' of production of the assets was generally lower than that which was necessary and that many new designs had paid insufficient attention to their needs for maintenance.

The 'quality' aspect became recognized at an early stage and efforts to enhance the management of quality became widely advocated and adopted. These initiatives have been refined and are continuing still today. Standards setting down guidance on the management of quality have been established both nationally and in the UK, (BS 5750) and internationally (ISO 9000 and EN 29000).

Similar attention to the aspects of maintenance and maintainability have not been given to the same degree. Maintainability and some features of maintenance support are now being addressed by the International Electrotechnical Commission (IEC), by agreement with the International Standards Organization (ISO), through their Technical Committee TC/56 'Dependability'. In addition the merits of 'preventive maintenance' as opposed to 'corrective maintenance' following actual breakdowns has been advocated on a fairly wide basis although there is little in the nature of standards yet.

In the immediate post-war years in the United Kingdom, it was principally those responsible for the management of the maintenance function who drew attention to the need to select assets, not so much on the basis of their initial cost but on the basis of their overall costs throughout their lives. Whilst the initial (capital) cost is a part of this cost of operating an asset, maintaining it throughout its life and finally disposing of it are equally important and can usually be far greater in value.

This plaint was being made increasingly through the 1960s and eventually was picked up by the (then) Department of Industry in about 1970 as one of a small number of developing industrial technologies. It was also felt, at that time, that it was desirable to develop an appropriate term for the identification of this concept of the economic management of physical assets, and this was referred to a panel of lexicographers. They recommended it be termed 'terotechnology', being derived from the Greek 'terein' which means 'to care for'. This was first published by the Oxford English Dictionary, in the 1986 supplement of the OED, although the first reference is given as 1970. It is used in equivalent forms in several other countries.

The term was used by the Department of Industry when they set up a Committee for Terotechnology in 1970. This committee reported in 1973. Recognizing that the principles involved were equally applicable for buildings, the Department for the Environment set up its own Committee for Building Terotechnology in 1974: this reported in 1977 under the title of 'Terotechnology in Buildings'. In 1975 a National Terotechnology Centre was set up to promote the concept; this had been recommended by the Committee but this was closed after a few years. Meanwhile the British Standards Institution was encouraged to extend its remit of an existing 'Maintenance' committee to cover the fuller scope of the terotechnology concept.

The definition of terotechnology was initially set by the Department of Industry's Committee and has remained virtually unchanged ever since. The current accepted definition, as laid down in British Standards is: "A combination of management, financial, engineering, building and other practices applied to physical assets in pursuit of economic life cycle costs". Two notes are usually appended to this definition to illustrate the intended scope. These are:

Note (1) Terotechnology is concerned with the specification and design for reliability and maintainability of physical assets such as plant, machinery, equipment, buildings and structures. The application of terotechnology also takes into account the process of installation, commissioning, operation, maintenance, modification and replacement. Decisions are influenced by feedback of information on design, performance and costs throughout the life cycle of a project.

Note (2) Terotechnology applies equally to both assets and products because the product of one organization often becomes the asset of another. Even if the product is a simple consumer item its design and customer appeal will benefit from terotechnology and this will reflect in improved market security for the producer.

Expressed simply, the concept promotes the application of *all* the necessary techniques to ensure that the user of the asset gets the very best possible value for his money. Whilst there are obvious advantages in the use of a single word to express such a concept, it is unfortunate that the term 'terotechnology' did not become generally accepted in the United

Kingdom. In this it differs from the term 'tribology' which was devised at about the same time and is now widely used. In part this was due to the term 'terotechnology', with its rather obscure root, not being one that lends itself to ready interpretation. It is therefore not surprising that it is understood by relatively few people in this country. The alternative title for the concept, as given in BS 3843, the British Standard 'Guide to Terotechnology', is 'the economic management of (physical) assets'. The adjective 'physical' is usually added to differentiate the assets concerned from the wider scope of this term as used by accountants and finance oriented practitioners. Although much longer, this alternative title does have the merit of clearly indicating the aim of the concept. For this reason the longer alternative has been used in the title of this book and is being advocated for use generally.

Associated technologies

The prime objective of the concept is the optimization of the economic well-being of the user of the asset. To obtain the maximum benefit for the user, all the techniques that can have a bearing on performance and costs should be applied. That is to say the full scope of terotechnology. Individual aspects of the concept are being promoted and practised in many organizations with varying degrees of success. However these are frequently known by different names. In some cases the technique, as originally promoted has been extended to cover a much wider range of applications until it has included a large proportion of the terotechnology concept. Retention of the original title of the basic technique has obscured both the full extent of terotechnology and the use of the term itself. The following list of practices, which is far from being exhaustive, are typical of those which adopt many of the principles of the concept and have become more familiar than the term terotechnology.

(a) resource management
(b) life cycle costing (LCC)
(c) through life costing
(d) total commitment for life
(e) womb to tomb costing
(f) cradle to grave management
(g) cost of ownership
(h) costs in use
(i) life long care
(j) design for life costing.

Of these many terms, 'life cycle costing' is probably becoming the most recognized and work is currently being undertaken by the IEC to produce a standard on this topic. This has arisen from a desire by certain user organizations, mostly in the public sector, to ask for life cycle cost estimates for their new assets and by their suppliers to furnish these in an agreed standard way. However, it is significant that it is proving very difficult to obtain international agreement on the scope of what is a life cycle cost. The main difficulty is determining whether a life cycle cost should or should not include consequential costs in the event of an asset breakdown. Whilst some organizations think such costs should be included others do not. Those arguing for these to be excluded are principally those who are aware that in their particular industries the consequential costs in the event of downtime can be very large, often several times the assets' capital value, even for a single incident, and may vary considerably according to the circumstances obtaining at the time of a failure incident. In these circumstances, the inclusion of consequential costs with their highly variable values is clearly impractical in a tendering situation. Furthermore, assessment of costs during the useful life phase of an asset are best carried out by the user organization rather than the supplier or manufacturer. The estimates are then more likely to take into account such confidential factors as the user's market expectations, which will be unknown to the supplier.

Assuming that agreement is eventually reached that LCC should not include these consequential costs, the LCC technique still has a major flaw. This prevents it from always establishing the optimum design or operational principles. This arises from its concentration on the costs, that is to say the outflow of expenditure by the user. Unlike a full economic management assessment it ignores both the extent and the timings of the inflow of finance from the sale of the asset's output. This could significantly shift the optimum situation for a commercial operation.

The correct management of an asset requires that at all times, the user organization is fully aware of the extent of all costs, past, present and future and also the returns from sale of its product, by-products and eventually from the disposal of the asset at the end of its life. Such assessments should commence immediately the project concerned is conceived. At that stage, the nature of the asset itself may not be known, only the product being proposed. Most of the costs used in these early assessments will have to be estimated. As the project develops and decisions are made on the asset's design, the basis of such estimates will be improved. By the end of the acquisition phase of the asset both its final design and its acquisition costs will be known accurately. At the same time, information

Table 1.2 Development of maintenance into terotechnology

Approx. Time of Progress	Earliest Days	Early to Mid 20th Century	Post WW2 c. 1950s	Later c. 1970s	Recent Mods, c. 1990s	Objective Fully Accepted Terotech.

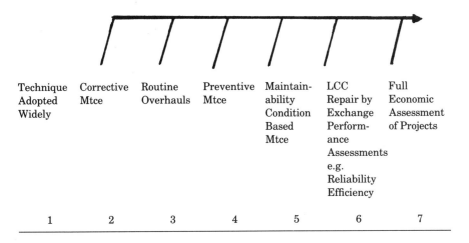

Technique Adopted Widely	Corrective Mtce	Routine Overhauls	Preventive Mtce	Maintain-ability Condition Based Mtce	LCC Repair by Exchange Perform-ance Assessments e.g. Reliability Efficiency	Full Economic Assessment of Projects
1	2	3	4	5	6	7

regarding the likely performance and operating and maintenance costs should be available from reliability, maintainability and maintenance support studies carried out during the development of the asset. Table 1.2 shows the development of maintenance philosophy and practices up to the introduction of the terotechnology concept.

Chapter 2

Organizations and Beneficiaries

As described in Chapter 1, the principal beneficiary from the practice of the economic management of assets is the user organization. When the user organization is large and sufficiently powerful to dictate its requirements to its potential suppliers, a greater proportion of the necessary activity to meet the objective falls to the user organization itself. This usually includes a definite selection of the product and the type of asset to acquire. Furthermore, the specification of the asset is likely to be set down in great detail, many options being pre-determined on the basis of experience of performance and costs in earlier projects.

Use of consultants

However, even large undertakings, such as government or a large public service operation, can seldom afford to employ specialists in every aspect of a project. Recourse to external specialist consultants for such aspects as are not covered by employees is the normal practice. This is a common practice especially for large infrastructure projects, civil works and buildings. In such cases it is essential that the project engineer or whoever is to control the project, ensures that the overall effect of the separate remits, such as civil works and buildings being handled by a consultant whilst the plant is handled by internal specialists, is one that achieves the optimum economic result overall. The author recalls a particular case where early commitment to an 'optimum' building and civil works left no room for standby plant which were subsequently shown to be desirable during plant development by reliability/maintainability studies. The ultimate result was a reduced asset availability and consequential losses of income significantly greater than the 'savings' in civil costs. That is to say a less than optimum design for the project as a whole.

It follows that a user organization employing external consultants should insist upon proper co-ordination with all other consultants and internal specialists to ensure that an optimum design, in terms of overall economics, is being achieved. As a corollary of this, consultants exhibiting a willingness for such co-operation and also a performance in their own work that shows beneficial economic influence on the project economics, should gain from the enhanced reputation these bring.

Benefits to consultants and suppliers

The case of a consultant benefiting from acting in such a way as to maximize the economic achievement of his client user organization is a special instance of a more general principle. This principle operates in all situations involving the contractual supply of a product or service to a customer. This is readily apparent in the case of a supplier of an asset to a user organization. If the supplier pays special attention to the user's needs for operability and maintainability, as well as performance factors such as efficiency and reliability, his reputation can be expected to improve. In turn the supplier's economic position can be expected to be enhanced through increased sales of his product or a greater market share. This must however be viewed correctly in the light of the user's economic performance. Improvements in a product's performance will invariably be achieved only at the expense of a higher initial cost. This should not be taken to an excessive degree such that the increased cost of the product, to become the user's asset, exceeds the economic advantage that the user is likely to derive from the improved performance.

In cases where a supplier/manufacturer has a product for sale to a wide range of customers, the design may have to be a compromise between the extremes of the customers' optima. Alternatively, two or more designs with different performance characteristics and costs may be made available. A supplier may have to choose between these possibilities and decisions made on the basis of the best economic advantage to himself. As part of this process, economic assessment of the range of users' life-time economics as well as the characteristics of the market will be essential. In other words it is necessary to carry out a series of economic assessments from the various users' point of view using feedback from the 'market'.

Relationship between supplier's product and user's asset

The second Note appended to the definition of terotechnology (see Chapter 1, under section on Historical Background) emphasizes the relationship between the *product* of one organization becoming the *asset* of another and the need for the user's requirements over a life-time to be considered.

Specialist buildings

Most assets will include a building or buildings. Where these are necessary to house the essential elements of the asset, namely the plant or equipment to be used to form the product, they are usually to a design dictated principally by that equipment. As a result they will normally be acquired in a similar manner to and directly in conjunction with the plant or equipment.

In the early years of this century it was quite common for associated buildings to be designed to have much longer lives than the plant they were to house. This was based upon an assumption that the original plant would, at the end of its life, be removed and the building re-equipped with new plant for a continuation of the project. However, more recent experience has shown this to be a false premise. Markets and/or methods of production are likely to have changed significantly requiring either different products altogether or greatly changed designs for such products. As a result the plant required to produce the product will have markedly changed and the original building(s) will not be suitable for housing it. This situation is aggravated by two further developing factors. Firstly, through technical advances there are likely to have been marked changes in the design of the asset plant or equipment or of the product itself even if this is nominally the same as originally produced. Consideration of the developments in domestic products amply illustrate this latter. Secondly, the size of the market for a given product may have changed so dramatically that the capacity of the production plant and hence the optimum sizes of production units could be so increased that the old buildings could make no sensible contribution to housing the second generation production plant. The electricity supply industry illustrates this. Just after World War II, the majority of new generators were of about 30 MW capacity. By the 1980s the industry was running several generators of 660 MW capacity. Realization of the effects of this led to a discontinuation of the financial practice of amortizing power station buildings over a much longer period than the plant it contained. As late as 1948 this method was being taught in text books as being the normal practice.

It follows therefore that buildings being sought for specialist plant and equipment should be designed with a life expectancy to match that of the plant or equipment it houses. Its design should be optimized such that its life-time costs should be a minimum taking into account all interactions in the overall economics of the project between the buildings and the rest of the asset. This will included any effects the building design might have on the operation and maintenance of the plant, or that these items might have on the maintenance of the building.

Non-specialist buildings

Specialist buildings of the type discussed above are in the minority. Buildings erected with the objectives that they be used for housing, offices or commercial premises are numerically more extensive. However, in such cases the relatively direct contracts between the user organization and its supplier(s) are seldom found. The more common arrangement is for the building to be devised by and designed for a developer, built by a contractor and finally sold to an 'owner' who, in most cases leases it to a user or a variety of users. With so many different parties involved, the link between the ultimate user and those deciding upon the details of the building's design and construction becomes extremely tenuous. Nevertheless, it will be to the user's advantage if the design and construction meets his needs in a way that is most economical for the conduct of his particular business. To a great extent he is limited in his possible actions to a choice between premises available to him on the open market. This is due to the extremely limited influence he can have on any particular premises and is generally limited to the individual internal fittings.

For the building owner, his economic advantage will mainly come from ensuring that the building will be leased as soon as it becomes available and remaining in this state for as great a period as possible during its subsequent life, thereby increasing his income. Depending upon the circumstances and his relationship with the developer he may be able to have some influence upon the design or construction which will make these objectives more possible or which reduces the likely burden on himself for maintenance. Acting not only in his own direct interests but from the point of view of potential users will undoubtedly help in ensuring the building remains leased to the maximum extent. He will also need to take into account not only the costs of maintenance but the likely costs of any modifications that may be required in the future or to attract future tenants and for this purpose the adaptability of the building design can be an influencing factor.

For the developer, his architects and the constructor a similar approach is desirable. In these cases the principal benefits are likely to arise from ensuring that their reputations and hence the likelihood of future commissions are enhanced.

Chapter 3

Sources of Assets

Turnkey contracts

The asset required for a given new project will normally comprise a number of constituents which may, indeed will usually, not be obtained from a single source. A 'turnkey' contract for an asset in which the total extent is to be provided by a single supplier organization tends to be the exception rather than the rule. Whilst such an arrangement can have contractual advantages, it is unlikely that the single nominal supplier can provide the complete asset totally from within his own organization and thus certain constituents have to be obtained from others via sub-contracts. This inevitably incurs the supplier in handling costs and can therefore add to the cost of the asset to the user.

Multiple direct contracts

The alternative of the user letting a series of direct contracts to all or most of the constituent supplying organizations may prove the cheaper option initially. However, it places upon the user organization the responsibility for total compatibility of the constituents and their matching when they are delivered to the site of the project. In the case of a complex asset, such as process plant or power station, this is an involved task with considerable practical as well as contractual risks. It can also require the setting up of a special department within the user organization to carry out these co-ordination tasks, or the employment of an external consultancy organization to carry these out. Either course will inevitably result in costs on the acquisition process which can partially or completely negate the apparent savings through separating the contracts for the several constituents.

The advantages and disadvantages of 'turnkey' and 'multiple direct' contracts are shown below.

Turnkey contracts

Advantages

(a) Co-ordination of components made the responsibility of main contractor
(b) Reduced size of user's project team
(c) Main contractor takes responsibility for sub-contractors

Disadvantages

(a) User has less control over designs of components and nominated suppliers
(b) User has less control over contract progress
(c) Profit margin added to costs (price) of co-ordination
(d) User has more difficulty in negotiating with sub-contractors

Multiple direct contracts

Advantages

(a) User has better knowledge of individual contract prices and can more readily negotiate on these
(b) User has more direct control over design of components and overall programme
(c) User has full control over co-ordination and ability to optimize the design of the asset
(d) Price of co-ordination is likely to be lower

Disadvantages

(a) User has need of larger project staff
(b) User has full responsibility for co-ordinating designs and programmes

Even in a case where an asset is relatively simple, say a small machine shop to produce motor parts, the asset will have constituents consisting on the one hand of production machinery and on the other the buildings to house the machinery, its operating personnel and those welfare facilities required by law. In such cases the arguments for acquisition by separate contracts for the different constituents will normally be stronger as the degree of necessary co-ordination is substantially reduced, albeit not entirely eliminated.

Early Phases

During the concept stage of a new project it may not be necessary to determine exactly which method of contracting for the asset constituents may be used. The uncertainties in estimates of both the initial costs and life-time costs should allow a general global estimate to be used for the costs of co-ordination of the constituents. This estimate should, of course, take due regard of the extent of the co-ordination task, based upon such experience as the user organization has of previous projects. However, a decision on the scope of the individual contracts, if a 'turnkey' approach is not favoured, must be made as the project progresses to the acquisition phase. This will have a fundamental effect on the various specifications and enquiries to be put to potential suppliers.

Acquisition by purchase

So far, in this chapter, reference has been merely made to acquiring the constituents of the assets and the placing of a turnkey contract or a series of contracts for their acquisition. In the case of a turnkey contract, the asset is normally purchased by the user using finance he has available or has acquired from a suitable source. The same method, straightforward purchase, can, of course, be used for any constituent separately contracted for. However, assets may be obtained in other ways.

Diverted assets or components

In the first instance, the user may already have the required asset or at least a significant component of it. If this is already or is shortly to become redundant for the purpose for which it was originally obtained, it might be redeployed to the new project. The original purpose will normally have been in respect to an earlier project, possibly one that is to be superseded by the new one. The component or asset concerned must be considered very carefully before being adopted and transferred into the new project. One of the principal factors must be whether it is in such a condition that it can be expected to be functionally serviceable for the life of the new project, or can be refurbished to such a standard at reasonable cost. That is to say reasonable when compared with the cost of an equivalent new item. This comparison of costs between the surplus original and a new item should not be made solely on the basis of initial expenditure but on the overall costs over the life-time of the project. With the pace of development of technology now being experienced, a new model may well have improved efficiency and/or lower maintenance or running costs which could outweigh the apparent savings through the use of an item designed and provided perhaps a decade earlier. A full economic comparison of the alternative courses of action should therefore be made. This should include due regard to any changes in ecological or statutory requirements that may have occurred since the original item was acquired, or are likely during the life of the new project. The factors and methods of carrying out these and other comparison exercises will be discussed in detail later in this book.

The ability to utilize surplus or redundant existing assets will generally be extremely limited where plant and equipment items are concerned. Apart from the rapid developments in technology which make such items out of date after only a few years, if they had been specified and optimized for the earlier project, their residual lives might well be expected to be limited, probably too limited for the new project. In addition, capacities of both project and the assets and components thereof have, in many cases, increased. This has arisen from business rationalizations as well as increases in demands and improvements in design and manufacturing capabilities.

Where simple machinery is involved, developments may have been very limited and in such cases a transfer to a new project is more likely to be possible. This will be especially the case where the nature of the equipment is, of necessity, robust, such as in a large vertical planer.

Re-use of buildings

One component of an original asset, that may well be suitable for adoption in a new project, is a building. With a reasonable maintenance, many buildings can have extended lives and remain serviceable long after their initial purpose has terminated. Adaption to a new project may well then be a viable option. This will be particularly true if the building is essentially a weather resistant shelter, such as a store or open workshop with few services. On the other hand, a purpose built structure like a hospital will have large numbers of built in services that would need to be completely replaced and significantly added to, for it to serve any new project. In addition, the methods of operating a modern hospital are quite different from those used half a century ago. To accommodate these changed methods, major changes to the fabric would be necessary and overall the economic choice is likely to be to demolish the old building and to rebuild.

Influence of timing

A factor, which is implicit in the comparisons discussed above, it that of timing. Today, great stress is properly put on getting a new product to market early. This urgency can influence the decision process in either of two opposite ways. Immediate availability of a redundant component, whilst a new component may have a long lead time for acquisition, may have a significant impact on the ready for market date. This is especially relevant where the supply of that component is on the critical path for the project. On the other hand, waiting for a component to become redundant, followed by a refurbishment of long duration may have precisely the opposite effect. The author has experienced cases where the refurbishment of an old component has taken significantly longer than the manufacture and supply of a new item, albeit at marginally less cost.

Should a new project be to produce a product already on the market but for which extra production capacity is required in the future, there may not be the same urgency to commission the new project. In such a case adequate planning time may be available and the prospect of using displaced components is enhanced.

Ownership

The two acquisition routes described above have one feature in common; this is that the asset or component concerned becomes totally owned by the user. As a result the user becomes totally responsible for all aspects of it throughout the working life of the project. This would include any modifications and, at the end of its life, its ultimate disposal as well as its operation and maintenance.

However, a project does not necessarily depend upon the user owning the asset or any particular component thereof. Items may be leased, hired or, in the case of buildings, rented. There are several advantages in acquiring assets by these methods. In the first place, it may reduce the initial capital expenditure on the project. This, in the case of a user organization with stringent limits on its ability to raise the finance for the project, could bring the project into the realms of the financially possible. Again, if the project's

Table 3.1   Summary of sources for asset

Item	Notes
Internal	
(a) Items already held by user organization and available and suitable for re-use	
(b) Items already held but diversion from existing use is economic possibility	Effects on new and donor projects must be established and evaluated
External	
(a) Straightforward purchase from supplier of items available for immediate sale	
(b) Contract with supplier for design, manufacture and supply of special items and direct payment for these under the contract	This would cover both special plant and equipment for the project and buildings
(c) Leasing through a separate financing organization	Generally used for plant which can be relocated and used when no longer needed for the project
(d) Hire of items, normally items needed for only a short time in the project	e.g. Construction plant and items needed for commissioning or periodically, such as for maintenance/overhauls
(e) Renting of buildings	Applicable when period of need is significantly less than building life and building is suitable for re-use Also used if initial (capital) expenditure has to be reduced or minimized

needs for a particular item is limited in time, such that only a small portion of its useful life is used, leasing or hiring may be appropriate. Such items as plant needed for the initial commissioning tests could fall into this category. Another example of this is likely to be in the case of buildings, especially those which have been designed to be flexible in use or are located in a position where they can be released to a further user and used in another project.

Another advantage of leasing such items, where this can be done, is that the user is relieved of the responsibility for disposal of the item and most of the administrative burden associated with this activity.

In the more general case of items being leased or rented, the responsibilities for modifications and maintenance generally will be determined in the contractual arrangements between the lessor and the user. Invariably this will be shared between the two parties in a way they mutually agree is equitable. The costs of the lease should reflect the extent of the lessor's responsibilities.

As a general rule the life-time costs of leasing an item can be expected to be somewhat higher than the corresponding direct costs had the item been purchased. However, the effects of indirect costs involved in its acquisition, use and disposal may outweigh the apparent difference and any effects of taxation rules must also be taken into account. The reduction in the initial, 'capital' expenditure could well be the determining factor for the user organization, even if the estimated life cycle costs of leasing were higher than for purchasing.

Table 3.1 shows a summary of the possible sources for an asset.

Chapter 4

Types of Organization and Management Principles

The economic management of its assets should naturally be carried out by any organization which owns or uses any plant, equipment, building or combinations thereof. In the extreme this should apply to an individual and his or her possessions. Predominantly though, we are concerned in this book with such organizations as companies, governments and councils at national or local levels and, for want of a better general word, institutions. The principles outlined will however apply in the case of individuals.

Objectives

In an earlier chapter we gave the simplified statement that the objectives of the management activity should be, for a commercial organization, the maximizing of its profits and, for a public service the minimizing of its costs. This is somewhat an over-simplification. For many commercial organizations the achievement of an improved profit may well be derived, at least in part, from the reduction of its internal costs to a minimum. On the other hand, government owned public service organizations may be set up and give a service on a commercial basis. The nationalized electricity supply boards that existed until recently in the United Kingdom are an example of such public owned commercial organizations, and many more such examples abound throughout the world. Such organizations are usually instructed to 'break even' over a period of years; that is to say to equalize their income with their costs. Even so they normally will be seeking to make a profit on the activities in any one year to give a margin against possible deficits in later years.

The essence of proper economic management of assets is based upon the use of appropriate economic assessments in all decision making. Whilst the costs of supplying the product are a major feature of these assessments, the organization's income from so doing is equally important. The way or ways in which this income is derived is therefore an important feature in the assessments and in categorizing an organization. The following case studies will illustrate the extremes that are encountered.

Case Study 1 – Coffee grinders

Firstly, let us consider a manufacturer of a domestic product, such as coffee grinders. Here the manufacturer will derive an income from the sale of his product which is more or less directly proportional to the number of units of product that he sells. If he has judged his project correctly and he can sell all that he can produce, this will accordingly be proportional to the availability factor[1] of his asset used to produce this product. If for market or any other reason he can not sell all the product he can make, he will deliberately reduce his utilization of the asset and output and income become proportional to the product of the availability and utilization factors. Such an organization is typical of what is referred to as a 'commercial organization' and will, of course, have as its objective the maximization of its profits.

Case study 2 – Public services

At the other end of the organization spectrum would be a governmental department or agency with a functional output, not readily measurable in any quantifiable manner as to extent or value. An advice bureau might be

---

1 In this context the 'availability factor' is one that takes into account reductions in capability as described in BS 4778 Part 3 Section 1. For example, if a plant is capable of running at full (100 percent) output for 80 percent of the time and at 50 percent output for 10 percent of time (and shut down for maintenance for a further 10 percent of time), an availability factor of 85 percent would be deemed to apply as shown in the following equations

$$\text{Availability is defined as} \frac{\text{'actual product produced in given time'}}{\text{total production possible in same time}}$$

Let us call 100% production in the period of time considered = Q then

$$\text{availability} = [0.8 \times Q + 0.1 \times 0.5Q + 0.1 \times 0Q]/Q$$
$$= 0.85Q/Q = 0.85 \text{ or } 85\%$$

an example of this. A local government's refuse collection service might be a further example of this. Its assets would include the collection vehicles, its fleet depot and its transfer stations. For this project the output would be fixed in terms of the collection and disposal of the local area's refuse. Clearly it should be the local management's duty to minimize the costs of this activity overall. Its income will in effect be the payment, as costs arise, by the local government from its taxation of the local population. The project activity must be carried out in such a way as to maintain the established performance requirement, which might be the regular collection of refuse from each of the premises in the area. Any proposed alteration to the established practice, in order to reduce costs, which would prevent this performance level being maintained, would, of course, be unacceptable.

Case study 3 – Income generating services

Between these two extremes there are many different types of organization. For example, telephone companies, which may be either publicly or privately owned, are one type of such organizations. In this type, its product is a service to its customers. But such a service is not completely quantifiable in terms of value to its customers nor in terms of the cost to the telephone company in respect of one customer's use of the system, the elemental basis on which the company's income is determined. As a result, the charges to customers, the source of the organization's income, are made on the basis of arbitrary formulae which are generally acceptable as reasonably equitable. Certain costs which can be directly attributable to a given customer, such as his connection and his terminal equipment can be and often is charged directly to him in the form of a rental. However, the bulk of the remaining costs, mostly associated with the capital and running costs of establishing and maintaining the network and its equipment can not be directly associated with either the numbers of customers or a measure of their utilization. There will be some association between the size and hence costs of elements of the network in terms of the traffic it has to handle and traffic must be a factor combining numbers of customers and their utilizations. But it also involves the company's own decision on an appropriate level by which demands will be met. In these circumstances it will be seen that any project to modify or extend the system will be very difficult to assess due to the problem of estimating the changes in income resulting from it. At best, based upon experience with other earlier projects, it might be possible to establish some approximate, empirical values that might be used in terms of changes in the numbers of customers, utilizations and the proposed levels of satisfying demands.

Facilities shared between projects

Finally, it must be remembered that supplier organizations such as manufacturers will also be user organizations as far as their plant, equipment, stores and factories are concerned.

At any given time most organizations will have a number of projects running and others in their preliminary stages. Again, in most cases, a given organization will have projects with many differing products. As each project is initiated, its economic assessments should be carried out on it as if it were in total isolation from the organization's other projects. This is, of course, an ideal and in practice there will be some material interplay with other projects. Sharing the marketing organization is a natural common feature. In such a case, in order to separate the assessments for the different projects, any costs involved, which need to be taken into account in the assessments, should be shared between the projects and the products in an appropriate manner. In other cases a by-product or waste from one project may form an essential input into a second project. In this sort of case the diverted material should be costed at an appropriate value in both projects. This value should be based upon the valuation that it would have if it were marketed separately instead of being diverted to the second project.

Financial rules

The financial rules for the economic assessments of a project should be set by higher management of the organization for that project. These may differ from project to project according to their nature and the circumstances in which they are being proposed. It is to be expected that the rules for a new major product, in the mainstream of the organization's objectives, would differ from those for a minor project. These may include such minor projects as one for turning a waste product into a saleable commodity or for establishing a stock of strategic spares for the production asset. Such schemes, whilst desirable, are not usually essential and are often classified as 'optional'. The rules concerned are those relating to the extent of finance available, the treatment of risk and the discount rate to be used in assessments. This latter factor could well be influenced by the source of the finance that has to be used.

However, where comparable courses of alternative actions are concerned, the same financial rules should always be applied. Decisions on such matters should always be made by the organization's higher management as they are fundamentally based on the organization's policy and the direction in which it intends to proceed.

We discussed earlier the impact of one project on another when material was interchanged or where some services were to be shared. However, there are two further ways in which one project can impact on another. In the first, the product of one proposed project might supersede the product of an existing project on the market. It might do this in terms of improved product specification, price, customer appeal or by some other means. In these cases, the assessment of the new project should take into account the effects of a revised assessment of the earlier project in the presence of the new project. This could lead to a modification, deferment or even abandonment of one or other of the projects in the best interests of the organization. Figure 4.1 shows the relationship between projects and the factors which need to be considered in the economic assessments of a new project.

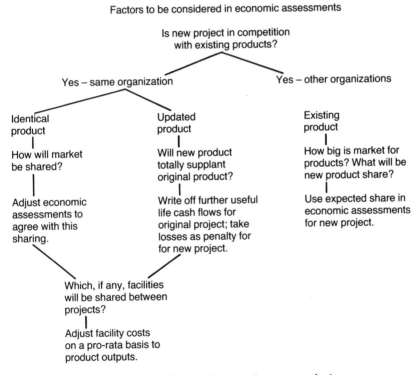

Fig. 4.1   Mutual influences between projects

Impacts on other products

Less obviously and far more difficult to assess can be the influence one product of the organization can have on another in terms of public response and its effects on marketability. If one product sustains a reputation of undesirability due to poor performance or the public's changed views on ecology, green issues or similar, it could affect the marketability of a further product, even one free from such disadvantages. The existence of such problems or the possibility of such problems arising in the future should be taken into account as far as possible in assessments, as well as guiding the organization's higher management on an appropriate response in terms of changes in policy.

A more direct impact on either product or the use of the asset can arise from changes in, or the introduction of new legislation or regulations. These can also affect the inputs to the asset in the form of labour or feed materials as well. More recently, regulations have incorporated reference standards into their requirements especially where European markets are concerned. Thus an awareness of the standards programmes, if not direct participation in the drafting of standards, should be undertaken as part of the technical assessment of a project. This is in addition to the need to anticipate the likely changes to legislation or regulations that are likely to occur during the life of the project.

Chapter 5

Life

The Project life

As the features of a new project are being developed during the initial stages of its conception, one of the principal aspects that must be determined is its proposed duration. From the point of view of the economic management of the assets to be used in the project, this is the principal 'life' that has to be considered. As far as the economic assessments are concerned the 'project life' extends from the initial concept until the assets involved are finally disposed of and the organization's responsibility for them has ceased. The 'project life' therefore starts with the initial concept and then spans the stages of its evaluation and authorization, followed by final product design, design and development of the required assets for producing the product, their acquisition and commissioning, their operation and maintenance and finally their disposal.

Asset and component lives

It will be obvious from the above that the actual life of the asset(s), which we shall call the 'asset life', extends only through the later stages of this scope and its working life, which usually determines the period for which it can be sustained, is somewhat less than its 'life' in terms of actual calendar existence. Furthermore, it may be found that certain components of the asset may have individual lives less than that of the asset as a whole. For such components the project has to be framed around the need to replace these, at the end of their working lives, by replacement components to enable the project to run to the end of the planned 'project life'. The components planned to be replaced at intervals during the project life will obviously have a deleterious effect on the asset performance whilst components are changed over, reducing the output and, where relevant, income to the project. As a result, whenever possible components should be designed to last for the full 'project life'. However, there will be circumstances in which designing a component to such a standard can be technically impossible or highly undesirable because of its impact on the design of other components or the asset as a whole. In a few cases it will be economically preferable to design a component for a more limited life, notwithstanding the economic losses of reduced production during replacement exercises. Such components as these will be referred to as 'life limited components', whether the limit is due to technical or economic reasons. As a general rule, the economically optimum situation will be achieved when the working lives of these 'life limited components' are designed to be such as to require only a very small number, usually the minimum practicable number, of replacement exchanges during the project life. Ideally, following the final exchange, the last component will itself become life exhausted at the final conclusion of the project.

Life phases

To achieve an economic optimum performance for a project, many different techniques should be applied in its management. These will be discussed in greater detail later in this book. However, different techniques tend to come into prominence during specific periods of the project life. For convenience therefore the project life is divided into a series of consecutive 'phases' in considering its economic management. These phases and their extents will now be outlined in the paragraphs below.

The Concept Phase

This phase is initiated when the first outlines of a project are expressed. Often this is limited to the identification of the proposed product. After consideration of the potential market, the likely demand for such a product in terms of quantities and the duration of the market will be estimated. The next stage in this phase is to establish the possible processes to meet this estimated demand and the assets needed to achieve this. Possible alternative methods may be identified. For each such alternative, estimates have to be made of the various life-time costs and the likely performance of the assets over the life-time in terms of output and any consequential incomes and their timings. The result of these studies should identify the best economic choice of the alternatives, which is then chosen as the basis of proceeding with the project proposal. Refinements of the details of this proposal and the cost and income estimates should continue until the proposal is in a state in which it can be submitted to the organization's higher management for authorization to proceed. In some organizations this form of the submitted proposal is sometimes referred to as a 'scheme'. At this point, the project proposal has to be considered and compared with other similarly proposed projects. In addition, at this stage, the higher management may have to consider their ability to finance the project. Their final authorization to proceed effectively ends this phase of the project life cycle.

The Acquisition Phase

This phase commences with the authorization to proceed which is usually accompanied by certain qualifications. These often include the requirement to seek further authorization from higher management before commitments are made for significant expenditure. At this time it is normal for the control of the project to be placed in the hands of a nominated person usually termed the 'project engineer or manager'. This phase is the period in which the 'project manager' remains in sole charge of the project and the term 'project' is often used in relation to this acquisition activity alone. This limited use of the term 'project' is common but it is important that it is appreciated that this is a very different interpretation of the term from that used in the economic management and assessments of assets and in the greater part of this book.

The first major activity in this phase is the preparation of the specification(s) for the asset and the issue of these, together with appropriate enquiries to potential suppliers. This is followed by the appropriate contract or purchasing procedures after the best supplier(s) have been selected. Too often suppliers are selected on the basis of the lowest tendered cost rather than the best economic tender. This latter should be determined with full regard to the effects of differences in design on the life-time costs and performance of the project. When the offered designs are significantly different this can require a very detailed examination of the likely life-time requirements for labour, maintenance, energy and all the other inputs to the process, as well as the likely performance.

After the 'best' suppliers are selected and contracts or orders placed, the supplier will commence the necessary design and development activities for the asset. In most major contracts the final proposed designs may have to be approved by the project engineer appointed by the user organization. Manufacture and erection or construction of the asset follows. Finally, the asset has to be commissioned: this activity may include, as required by the contract, appropriate performance tests. Upon satisfactory completion of the assets and these tests, the user takes over the asset. During the later stages of this phase there will be coincident activities associated with the training of the user's operating and maintenance personnel and the delivery of all the required documentation for the asset. The final taking over of the asset by the user concludes this phase.

The Useful Life Phase

This is the phase in which the asset is put to work and the product produced. It usually extends for the greater part of the total project life. Two activities predominate in this phase, the operation of the asset and its maintenance. The activities in this phase are therefore normally divided into these two aspects. Operational aspects include the planning and execution of the operation of the asset in a safe manner and at minimum possible cost per unit of product, consistent with meeting the demand for it. This will involve a critical examination of all aspects of the operation of the asset components to minimize costs and maximize output of the product. At the same time maintenance is carried out to keep the asset in an effective state for producing the product. The types and extents of maintenance have to be judged to minimize the consequential total cost of the maintenance activity. In this context the need to shut down the asset or otherwise reduce the output of the product has to be taken into consideration as well as the direct costs in terms of effort and materials in carrying out the mainten- ance. Furthermore, a similar balance has to be made between the direct consequences of failures of the asset or its components and the mainten- ance effort aimed at preventing or minimizing such failures.

The operational and maintenance aspects cannot be viewed in total isolation as much as maintenance activity is bound to impinge upon the operation of the asset and the output of the product. Conversely, mainten- ance is invariably aimed directly or indirectly upon the asset's continuing ability to produce the product in a safe and efficient and economic manner. Safety is invariably the predominant factor affecting both operational and maintenance aspects. If safety of personnel or the general public is obvi- ously infringed, even if no injuries have resulted, losses of production and penalties under legislation or regulations are likely to be so severe as to overwhelm the costs of carrying out a more responsible course of actions. In all these aspects the co-ordinated planning of operational and maintenance activities, to achieve the best economic performance overall, becomes a major task in the useful life phase.

Throughout the life of the project and especially during the useful life phase, periodic reviews should be carried out into the economic perform-ance of the project both to date and as expected at its projected termination. As the project approaches its originally planned termination date, these reviews should additionally examine what options exist after that date and to assess their economic implications. If there remains a demand for the product and the asset still retains a capability for producing it, either with or without a degree of refurbishment, there exists a possibility of extending the originally planned project life-time and deferring the termination date. The economics of this, in absolute terms and also against the economics of a new supplanting project must be the determining factor on whether an extension is justified. It will be appreciated that these economic assess-ments of the project will depend upon a detailed technical assessment of the asset and predictions of its future performance based upon this.

The Disposal Phase

When the termination date is reached, either the originally planned or the extended date, production will have ceased and the disposal phase is entered. The first action to be taken by the user will be to render the asset safe. That is to say safe both as a standing entity and also safe during the disposal options. The possible options must be considered both for the asset as a whole and also for its several components. The course of action for each of these items must be evaluated and finally selected on the basis of which combination gives the best overall result in terms of possible income or expenditure for the user organization. Some of the possible courses of action are:

(a) Sale of the complete asset as a going concern.
(b) Sale of components individually for use by either other organizations or the transfer of these into other projects of the user organization.
(c) Sale as scrap materials.
(d) Disposal as unwanted materials.

Combinations of these courses (b) to (d) is also possible and often likely. In addition, courses (b), (c), and (d) will involve dismantling the asset and this will inevitably involve costs either directly or in terms of a reduced price for the sale of the component.

Sale of the complete asset as a going concern may be possible when technical assessments have shown that continuing operation is feasible but the original user has declined to extend the terminal date for economic or other reasons. This will be more often experienced where a building is concerned. This course of action has the merits of providing further income for the project and at the same time terminating almost all the user's responsibilities for the asset. Only responsibilities emanating from the contract for the sale would remain.

Course of action (b) also has the merit of producing some income for the project and, for the components concerned, a similar reduction in future responsibility.

Course of action (c) will usually be less desirable as the levels of remuneration are likely to be significantly less than for course (b).

Unwanted materials will eventually have to be disposed of as given in course (d). Inevitably this involves the user in costs both for the taking away and then for storage. Where undesirable materials are concerned, either toxic materials like certain chemicals or asbestos, environmentally deleterious materials such as PCBs, or radioactive materials such as from nuclear power stations, these disposal and storage costs can be very significant and could have a noticeable affect on the overall project economics.

When an asset contains some 'life limited components' some of such components will be taken out of service and replaced in the asset, at intervals during the useful life. As a result the economic disposal of these has to be undertaken or at least considered at those times. Because, by definition, these will have exhausted most, if not virtually all of their working lives, the options for their disposal will be limited. The course of action chosen should be that which is economically best for the user, preferably their sale for the highest price available. Although these components may have had the majority of their working lives exhausted, in the context in which they have been employed, before being replaced, they may still be usable in a different context. In such cases they may have a value to another user and can be sold at a higher price than they would achieve as material scrap. The possibilities of such a course of action are therefore well worth exploring and adopting if at all possible.

If no such possibilities exist, the sale of these exchanged components as scrap material may remain as the best economic option. Should even this prove to be not possible, the components will be destined for eventual disposal and consequentially a cost upon the user. In this final event, consideration might be given to the merits of storing them until the end of the project and then their disposal along with the finally exchanged items and other components destined for similar disposal. Any costs of storing these components until the end of the project life must be weighed against a

Table 5.1 Comparison of life-time phases and stages as used in the three managerial disciplines

Economic Management	Project Management	Dependability
Concept Phase	Conception Phase Feasibility Phase	
Acquisition Phase	Implementation Phase	Design & Development
		Production
		Installation & Commissioning
Useful Life Phase (Operation & Maintenance)	Operational Phase	Function & Maintenance
Disposal Phase	Termination Phase	

reduced cost of disposing of all such items in a single action. This might be worth while if there were special constraints upon such disposals as there could be if ecological considerations or radioactivity were involved. Again economic evaluations of the various possible courses of action should be carried out and the results used as a guide to the choice of course of action to be taken. Should doses to personnel of radiation or other noxious substances be involved, a similar evaluation in terms of dosages to personnel should be carried out and the course of action to be taken may well be taken on the basis of minimum dosages rather than minimum costs. Legislation or statutory regulations may insist upon such a course of action being taken. As well as to life limited components, similar considerations may apply to components to be disposed of in the final disposal phase.

In the economic management of assets, matters are invariably viewed from the point of view of user organization. It is for this reason that the project life-time is viewed as consisting of the four phases – Concept, Acquisition, Useful Life and Disposal in which distinct optimization techniques are mainly applicable. Other disciplines, with different viewpoints tend to look at the project life-time in slightly different ways and with slightly different temporal divisions. Two such viewpoints are those of the reliability/dependability specialist, which views things mainly from the supplier's point of view, and for the project management specialist which takes either of these viewpoints. The project life as broken down from these points of view are given respectively in BS 5760 and BS 6079. These are illustrated in Table 5.1 which shows the relationships between the phases from these various viewpoints.

Summary

Table 5.2 summarises the principal activities towards the economic management of the asset throughout the project life.

Table 5.2 Principal activities during asset life-time

Phase	Stage	Activity
Concept		Devise outline of project including basic design of asset. Investigate need for product. Decide capacity of project. Estimate capital and running costs and value of income. Consult other departments for requirements and conditions. Prepare and submit proposal & economic assessment for authorization.
Acquisition	Preliminary	Appoint Project Manager. Decide contract strategy; assess potential suppliers. Prepare Specification(s) with inputs from specialist departments. Issue Enquiries.
	Contract	Preliminary assessment of tenders. Clarification of tenders before selection of contractor(s). Economic assessment to check for viability. Place contract(s).
	Design/ Development	Design asset. Carry out Functional, Value, Reliability, Maintainability assessments and refine design. Submit design proposals as agreed.
	Manufacture	Plan manufacture and commissioning programme. Manufacture and test. Monitor progress against programme.
	Delivery Commissioning	Carry out erection and installation procedures. Test components and systems. Test overall system and commence test production/operation. Check performance and output against specification and guarantees.
Useful Life		Commence operation and refine procedures. Commence maintenance – both corrective and to maintenance plan Initiate plant and operational histories – analyse performance and cost data for comparison with published data and record for use in future projects.
Disposal		Assess prospects for further use of asset or components; check need for use in own organization. Check potential for resale. Devise disposal strategies to give maximum income consistent with environmental requirements. Carry out disposal plan.

Chapter 6

Composition of the Asset

General

In this book, devoted as it is to the economic management of projects, the term 'asset', in the singular, will normally be used for the collective totality of the components required to produce a given project's output of product. This use of the singular form is to emphasize the essentiality to consider each and every one of these components as a contributor to the project's objective and to be included in its assessments. When a project is to produce a material output, whether this be a manufactured item or a commodity to be handled in bulk, such as soda crystals or petroleum, the principal component of the asset will be the plant needed to carry out the processing of feed materials into the finished product. However, both the plant used for the processing and the personnel who operate it will need protection from the elements and so will normally be housed in a building. Such a building, if devoted solely or principally to this purpose is clearly essential to the project and must therefore be considered in economic assessments as a component of the asset. Two examples of the extent of an asset are shown in the following two case studies.

Examples of the composition of the asset

Case 1

This example illustrates the extent of the total asset for a relatively small and simple project. The project is to manufacture, in quantity, a domestic appliance, say a vacuum cleaner, in a purpose built factory.

 (1) The tools and machinery to fabricate the product

  (2) Assembly benches
  (3) Store for incoming feed materials – plate metal, fasteners, electric motors etc.
  (4) Store for completed appliances and packing facilities
  (5) The factory building together with welfare facilities required by law
  (6) Test bay for products and quality assurance inspections
  (7) Drawing office
  (8) Lifting and handling equipment for factory and stores
  (9) Technical manuals for all plant, machinery and buildings
(10) Maintenance workshops/tools
(11) Production/sales office (possibly shared with other projects)

Case 2

Taken from the author's experience, this example is of a project to hold a strategic spare rotor for a population of high pressure steam turbines.

  (1) The rotor itself, fully bladed
  (2) A storage/transport stand to keep the rotor in (with axial chocks to restrain rotor whilst in transit)
  (3) A zipped plastic cover to keep the rotor in whilst in the stand
  (4) Mesh covers to afford physical protection whilst in transit
  (5) A dehumidifier to circulate dry air in the plastic cover whilst in store
  (6) Spare bearing shells matched to spare rotor journals
  (7) A complete set of gland segments for interstage and casing glands (to expedite fitting or to replace glands damaged in incident causing rotor replacement)
  (8) Schedule of complete fitting/clearance dimensions: identification of any non-standard parts (e.g. blade roots)
  (9) Data needed to fit rotor to all machines of family
(10) Any special parts needed to fit rotor to any non-standard machine (usually caused by manufacturing error compensated by fitting non-standard matching pieces)
(11) Oversized coupling patch plates and complete set of coupling bolts (to be finish machined to fit reamed coupling bolt holes)

Storage facilities

The total process involved in the project may require storage of either the feed materials to the process or of the finished product from the process. In some cases, storage of part finished material between stages of the process may also be needed although optimization techniques such as JIT are tending to minimize this. Such storages and any equipment they may contain for handling, loading or transporting should be included in 'the asset'.

Items shared between projects

Where buildings etc. are used in more than a single project, as far as costs go, they must be proportioned in an equitable manner between the projects involved as described in Chapter 5.

Services and the plant associated with them may also be either devoted to a single project or shared between a number of projects. Again the costs of these, proportioned if appropriate, should be included in the scope of the asset.

Maintenance equipment

Equipment used for the maintenance of all the components of the asset must also be included in the totality of the asset, when this is used solely for one project. As in the case of buildings and services, if a piece of maintenance equipment is shared between two or more projects, a proportioning must be carried out. The two major items within this category are likely to be the workshops and their equipment used for maintenance of the asset and any special test equipment that may be required. General purpose tools and test equipment will normally be accumulated in terms of costs which would be allocated to the cost of maintenance but are unlikely to be formally included as part of the asset. That will be unless the aggregated cost is significant in terms of the total value of the asset and that they are expected to last for the total duration of the project and furthermore entirely devoted to the single project. Clearly this excludes small tools and minor equipment such as insulation test sets which are expected to have a limited life and to be replaced periodically.

Control equipment

Where an asset includes electronic boxes either as the principal component or as a feature of control mechanisms, not only test equipment but specially designed test benches and panels may be required. These may be needed for maintenance but equally may be required to enable the equipment to be set up to enable it to perform properly. Such items would be appropriately deemed to be included in the asset. Similarly software for any programmable devices must be considered in the same light.

Manuals

Finally, most plant and equipment, especially that incorporating recently developed technology, will require both operating and maintenance instructions. These should be included in a technical manual. This technical manual should ideally conform to BS 4884 but, in any case, should meet all the user's requirements for documentation. Thus in addition to the operating and maintenance instructions the manual should include, as required by the user, chapters giving general and technical descriptions, maintenance schedules, handling, storage and erection instructions and full safety information. Full drawings of the item as erected may be included in the manual; if not these will be needed in addition. This is especially true when the item is a building. All this information should form a component of the complete asset.

Training aids

The operational staff may not need, apart from the technical manual, much in the way of additional essential equipment. However, in the case of large technically advanced assets such as an airliner or a nuclear power station, operational staff may need to be trained and to practise on simulators. When these are necessary, usually for safety reasons or to practise emergency drills which can not be exercized on the asset itself, they must be deemed to be a further component of the asset.

It will be clear that the aspects discussed above must be fully considered before the complete extent of the asset, required for a given project, is determined. In the economic assessments, carried out during the project life, both the initial and running costs of all these components must be considered.

Chapter 7

Economic Evaluations

The need to carry out economic evaluations has been referred to many times in the earlier chapters of this book. This chapter outlines the bases on which these should be performed and some of the principal factors involved. Much of this may well be familiar to some readers. However, far too many engineers and other persons, who are involved in substantial projects, have never fully understood the proper basis for carrying out such exercises. This chapter is aimed principally at such initiates.

Periods of time to be considered

Most projects, including the simplest, involve project life-times extending for many years. Even in the domestic area, most housewives will normally expect their kitchen appliances, the cooker, dishwasher and washing machine to have reasonable working lives, of the order of 10 years or more. When we turn to major industrial or governmental projects much longer lives are expected. A warship will normally be expected to have an operational life of 20 to 30 years. Aircraft, both military and civilian airline types are normally expected to serve for similar periods. If we add the time taken to design and develop, manufacture and commission such assets, the project life-time could well be an extra five years on top of these figures. A probably extreme case exists in the case of a nuclear power station. Design and construction would be of the order of five years and its working life planned for 30 years. Thereafter decommissioning would take place in stages extending for a further 100 years until it is finally completed and a 'green field' site restored. It is therefore clear that both project costs and incomes, where relevant, will take place spread over a significant number of years, and this must be taken into account in the economic assessments.

Basis of evaluating cash flows

In economic evaluations, it is important to identify exactly when, in the project life-time, a particular cost is expended or a particular sum in the form of income, is received. This is due to the fact that the value, to the user organization, of such sums is dependent on the times that they occur. For any reader not already fully familiar with this concept, the following simplistic explanation is offered. Consider an investment being made of a sum of money 'A' in a bank or a similar institution. It is to be expected that this will attract interest and, if no withdrawals are made, this will accumulate in the form of compound interest. At some later date the investor can withdraw his investment, together with the accumulated interest, to receive a sum of money 'B' which is greater than 'A' by the amount of interest. It follows that, to the investor a sum of 'A' at the investment date is equivalent in value to a sum of money 'B' at the later withdrawal date, say in 'X' years' time. In common investment terminology this is expressed as the 'present worth' of a sum 'A' today is equivalent to the 'present worth' of a sum 'B' in 'X' years' time. In terms of value, this can be expressed as:

Value of 'A' today = 'B' × present worth factor in 'X' years' time.

From the explanation of how 'B' arises (from the addition of interest), it is clear that this 'present worth factor' must be

1/(1 + rate of interest received after 'X' years)

An alternative method of viewing this is as follows. If I have to spend a sum of 'B' in 'X' years' time, I only need to put aside a sum 'A', today and invest it, then draw it out, with its earned interest, in time to pay the sum 'B' when it becomes due.

This is a fundamental aspect in comparing the values of cash flow which take place at different points in time. In the explanation given above we referred to the rate of interest received on the investment. This is an identifiable rate in the example as most banks or similar bodies advertise in advance the rate of interest which will be given to investments. In economic assessments other factors may affect the situation other than 'bank interest' rates. So in such assessments we use instead a 'discount rate' rather than a 'bank interest rate'. We shall discuss how such 'discount rates' are derived later in this chapter. However, using the appropriate 'discount rate', determined for the project, we can use the above method to convert any future (or past) expenditure, (or income) to its equivalent 'present worth' at a nominal base date. This base date is usually at the start of the project, (in the concept phase) and we shall term, for this book, this as 'base date P'.

Inflation and variations in real prices

In addition to this fundamental aspect, there are other time-dependent aspects to be considered. When dealing with costs and income sums over an extended period such as a project life-time, two other factors have to be considered which affect the value of sums of money in the cash flow expressed in the value of the currency at the time they are exchanged.

Firstly, there is a tendency for the value of a currency to drop with time due to general inflation. As a first step therefore in establishing 'value', a sum of money expended or received at some given date must be adjusted in terms of a standard scale of values for currency. This is normally done in terms of a unit of currency at a given base date, usually at the start of the project. This date is usually chosen, as the value at this date is known and can readily be used as a reference throughout the project life. We shall call this 'base date C' for this book. The corrected cash flow, in terms of the currency at the base date, is normally expressed in terms of '£s at 19– (the start of the project, and "base date C")'. Colloquially this is often referred to as the 'real value'. The effect of changes in the value of the currency are normally related inversely to the general cost index. In the United Kingdom this is normally taken as the Retail Prices Index (RPI) for the year in question.

A very large number of commodities tend to vary in cost in direct sympathy with the general cost index over a protracted period of time, although there may be short term variations from this norm. For such commodities it is usually sufficient to correct their value, in terms of the currency of the time, to their value in 'real terms', namely their value in terms of the currency at the 'base date C', by using the general cost indices. These cost indices are usually published in tabular form by the government concerned. For costs, etc. likely to arise in the future, for such commodities, it is to be expected that these will vary with the general index (e.g. the RPI) which can not be predicted with any confidence. However, in terms of the currency at 'base date C', that is to say the 'real costs', such costs can be expected to remain sensibly constant.

Because of market conditions, certain costs may increase or decrease over time at a rate greater or less than that of general inflation, which governs the value of the currency. Such conditions can be expressed as 'costs rising (or falling) at a rate of y percent per year in "real" terms'.

Materials in short or limited supply, such as fossil fuels, are typical of those which tend to rise in 'real' terms, whilst the cost of manufactured products which are subject to extensive and continuous development, have tended to fall in recent years. A possible method of dealing with such commodities is as follows. For present and past expenditures the actual prices will be known, in terms of the currency at the time they were expended. These can be corrected to the currency basis by the use of the general cost indices as explained above. For future expenditures costs can be estimated by using the latest cost figures, in 'real' terms, and correcting these for the years in question by applying the expected variation rate (i.e. the 'y' percent per year mentioned above). This expected variation rate is derived from experience to date and should ideally be calculated by a person fully experienced in the market involved. Such calculations will show cash flows in the terms of the currency at base date C, as they are already based upon 'real' costs.

The methods described above allow us to convert actual and predicted future costs and incomes to a basis of the currency at base date C (i.e. 'real values'). They, however, remain as cash flows at the times when they have occurred. It is still necessary to put these into a common basis before they can be compared or subjected to arithmetical treatment. This is done by converting the cash flows, in 'real' terms at these various dates, to their 'present worths' as described at the start of this chapter. This process uses the 'discount rate', prescribed by the user organization, for the particular project and the 'base date P'. In most cases, a year at the very start of the project, in the concept phase, will be used for both 'base date C' and for 'base date P'. However, in order that the basis of cash flows is fully defined it is important that both base dates are stated in reference to any schedule of 'present worths' of cash flows for a project. This might be expressed in a common statement in association with such cash flows in a form such as 'present worth values in 19– in terms of £s in that year'. Convenience usually dictates that a common year is used for both base dates but it must always be remembered that this is not essential and so clear statements are necessary for both. To complete the statement, defining the basis of present worth values, it will also be necessary to state the 'discount rate' that has been used to correct future 'real values'.

Problems in media reporting

Many examples abound of technical papers and articles which quote figures for costs, in long term projects, without precisely defining the basis

Table 7.1　Examples of extracts from present worth tables

Discount rate (%p.a.):	5	10	15
Year			
0	1.0	1.0	1.0
1	0.952381	0.909091	0.869565
2	0.907029	0.826446	0.756144
3	0.863838	0.751315	0.657516
4	0.822702	0.683013	0.571753
5	0.783525	0.620921	0.497177
10	0.613913	0.385543	0.247185
20	0.376889	0.148644	0.046201

on which they have been assessed. Quite often a sum of money is quoted without a clear statement of whether this is in terms of currency at the time or in 'real terms'. Even when it is made clear that the sum is in 'real terms', the base date, that is 'base date C' will not be identified. Recent papers discussing comparative costs of producing electricity from gas turbine, fossil fired plant and nuclear generators have identified the 'discount rates' assumed in the estimates. However, full and clear statements of all the various factors, namely 'base date C', 'base date P' as well as the discount rate are seldom made. This makes it extremely difficult for interested readers to use the figures and to convert them into other bases which they may consider more relevant from their particular point of view. Poor as these technical publications are in this respect, the media and indeed the statements made by politicians, tend to be even worse.

The process described above of converting costs to a common basis of 'present worths' is described in many text books on accountancy. It is usually referred to as 'Discounted Cash Flow' or 'DCF'. Standard tables are available which give values of the 'present worth factors' for the various years after the 'base date P', (termed Year '0' in the tables), and for various 'discount rates'. These render the conversion of the 'real values' to 'present worths' a simple operation in practice. An example of such a table is given in Table 7.1.

In summary therefore, a given 'cash flow', (expenditure or income), in a given year can be expressed in either of three forms. These are (a) in terms of 'actual cash flow' (in the currency of the time), (b) in 'real terms' (in terms of the equivalent value of currency as 'base date C'), or (c) in its 'present worth' value (its equivalent in value at 'base date P'). The conversion from

(a) to (b) or vice versa is achieved by use of the general cost indices for the given year of the cash flow and for the year of 'base date C'. Similarly a conversion from (b) to (c) or vice versa is carried out using the 'present worth factor' for the year of the cash flow, relative to 'base date P' and the specified 'discount rate'. An example of how these equivalent 'costs' vary is given in Table 7.2.

These three different expressions of cash flow are used in different circumstances. The 'actual' value of cash flow is important to those dealing with the particular transaction at the time and is relevant in budgets and financial records which deal with matters in terms of the currency of the time. 'Real' values are of use when comparisons have to be made of cash flows with predictions made at an earlier date, possibly at the time of the original concept of the project. 'Present worths' are the only basis on which aggregations of costs and incomes can be assessed to determine the optimum way ahead, when alternatives are available or, looking at the project as a whole, to determine its viability. Conversions of these cash flow values, as required for the various purposes described above, is a simple matter using the methods described in the preceding paragraphs, provided the three factors 'discount rate', 'base date C' and 'base date P' have been properly established. An example of how values are aggregated is shown in Table 7.3.

Purposes of assessments

Economic assessments are essentially of two types according to their purpose.

(a) In the first type all the costs and incomes throughout the complete project life time must be taken into account. These are then summed (in their present worth forms) to determine the overall viability of the project. For a commercial venture, if the sum of all the incomes exceeds the sum of all the costs, the project will show an overall profit. In the case of a service project, say to provide a town with a refuse disposal service, there will be no incomes in the normal sense as costs are directly met by the local government. However, the aggregation of the present worths of the costs will give a yardstick whereby the viability of the project can be judged.

(b) The second type of economic assessment exercise is used when comparisons have to be made between alternative courses of action. At the concept stage, when a proposed project has to be submitted to higher management for sanction, it may be in competition with other proposed

Table 7.2 Equivalent costs

Year	Actual cost (Expenditure) £	'Real' value of cost (In Year 0 terms) £	Present worth of cost (In Year 0 values) £
0	100 000	100 000	100 000
4	10 000	7835	4865

In the example the 'real' costs are based upon inflation being at 5 percent per annum until Year 4 and a discount rate, based upon 'real' values, of 10 percent per annum is deemed appropriate. Figure 7.1 shows, over a longer period, this decline in 'real' and 'present worth' values.

Note: The present worth factor (r) for translating actual costs directly to present worth values in the example is given by

$$(1 + r) = (1 + 0.05) \times (1 + 0.10)$$

which gives r = 0.155 or 15.5%.

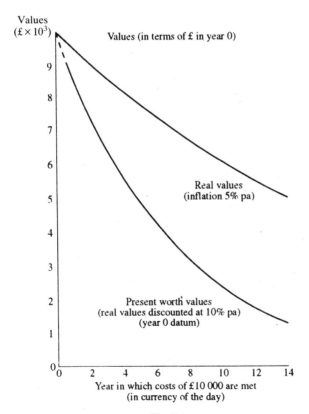

Values (in terms of £ in year 0)

Real values (inflation 5% pa)

Present worth values (real values discounted at 10% pa) (year 0 datum)

Year in which costs of £10 000 are met (in currency of the day)

Fig. 7.1

Table 7.3 Example of aggregating costs

Scenario: An asset costing £100 000 is purchased. Its costs in use over the ensuing five years of life is £20 000 per year for the first two years and £25 000 per year for the final three years. A discount rate of 10 percent per annum is appropriate for this project: inflation can be ignored.

Year	Actual costs	PW factors	PW of costs
		Expenditures (£)	
0	100 000	1.0	100 000
1	20 000	0.909	18 180
2	20 000	0.826	16 520
3	25 000	0.721	18 775
4	25 000	0.683	17 075
5	25 000	0.621	15 525
Totals	215 000		186 075

Note: Whilst the actual expenditure on the project amounts to £215 000, and this must be budgeted over the six years of the project life, the equivalent value in terms of an investment in Year 0 is only £186 075. This is the sum of money to be made available in Year 0, which, if invested in a bank at a 10 percent per annum return until actually needed to meet expenditure, would be sufficient for the project.

projects for limited financial resources. In these circumstances comparisons have to be made between them partially on the basis of their individual assessments of viability, as described above. However, in the majority of cases the comparisons that have to be made relate to a more limited impact on the cash flow within a single project. In such cases, the optimum course of action will be determined by the overall effects on the aggregate of the life-time cash flows. However, in these cases, many elements of the cash flows will remain the same for the different courses of action. Since these are common to all, these may be ignored and only those elements of cash flow which are different between the alternative courses of action need to be considered. The course of action with the lower aggregated net cost (i.e. costs less incomes) is the optimum choice. It should be noted that in these cases, differences in timing may be as significant as differences in the amounts of the cash flows.

Discount rates

We must now turn to the question of the 'discount rate' that should be used in these assessments. This is essentially a matter that should be determined by the senior management of the user organization on the basis of their needs. The following paragraphs are intended as an illustration of the factors which they should take into account in setting the discount rate to be used in the assessments of a given project.

Source of finance – banks

Let us first consider a simple project in which costs and incomes can be predicted with a high degree of confidence. Assume that this project is to be financed by a loan from a bank which has declared it requires to receive interest at a rate of 'X' percent per annum on the sum borrowed. For simplicity let us ignore any questions of grants and assume that allowance for taxation has been taken into account by including these in the schedule of cash flows. In this simple case, an element of cost 'A' will give rise to a borrowing of this sum against the loan and must be paid back with interest at a later date. Assume that this element can be repaid from income one year after it was borrowed. The sum to be repaid will be 'B' where this is equal to A × (1 + X/100). In equivalence terms therefore 'A' is equal to 'B', one year later. Mathematically, this is the same as using a discount factor of 'X' percent per annum: the use of this discount factor on future sums of cash flow will therefore make the necessary adjustment to ensure that any sum borrowed can be repaid with interest at a later date when income makes this possible. Overall the net sum of the costs less incomes at the end of the project will indicate whether the project is viable. A negative sum, indicating that the present worth of incomes exceeds that of costs, will indicate that a profit will be made.

Source of finance – internal

Let us now consider a similar simple project, but in this case the user organization already had adequate funds to cover the needs of the project. In this case the organization has to decide whether to authorize the project or to invest its money instead. Say an investment giving 'X' percent per annum interest in available. In the light of this alternative, the project will be expected to give a return equal to or greater than 'X' percent per annum. If the project is now assessed in total a discount rate of 'X' percent per annum should be used. This will ensure that when costs and incomes in real terms are discounted, the returns on money utilized in the project will be sufficient to pay back this money plus the necessary interest. This is, of course, assuming that the net present worth, at the end of the project life, shows it to have been profitable.

Source of finance – financial market

A third alternative for the simple project might be if the organization raises the finance for it by issuing shares and borrowing from investors. Assuming that the organization aims to pay dividends amounting to a return of 'X' percent per annum on the sums invested, the situation is similar to that of the last paragraph where the pay back is to match or exceed that of the alternative of investing the money. Again a discount rate of 'X' percent per annum would be appropriate to ensure that adequate returns are made.

Basis of establishing discount rate

In practice, an organization may raise the finance for a project in any one or any combination of these methods. Based upon which ways were used, a discount rate could be established which would ensure that the necessary returns were made on the sums used in the project. Again, in practice, this figure for the discount rate will have to be adjusted to take into account the effects of any grants or taxes. It may also need to be raised somewhat to allow a margin on the uncertainties that can arise, or are believed to exist, particularly those due to possibly over optimistic values of incomes or costs.

Changing quoted rates to 'real rates'

In the above examples, as was stated, it has been assumed that the cash flows had already been converted to their 'real' term equivalents, and consequently the discount rates were in these terms also. However, in practice, banks and other lenders invariably quote their terms for interest on loans in terms of the currency of the day and deal with actual cash flows on the same terms. Calculations of actual interest required are again in the terms of the currency at the date it was borrowed. In setting their interest rates the bank will have due regard to the fact that the value of the currency is falling as inflation has an effect. Thus although a bank may require an interest rate of 15 percent per annum, if inflation were progressing at a rate of 5 percent per annum in real terms, the true rate of interest (in real terms) would be less than 15 percent per annum as part of this value would be devoted to maintaining the purchasing power of the original sum borrowed. It follows therefore that a discount rate based upon real values, not the actual values of cash flows, should only be compared with the equivalent interest rate on borrowing or the rate of return required in the same 'real' terms. Alternatively, a higher discount rate can be used which can be directly compared with the interest rate and used operating on the actual cash flows rather than their 'real' values. Whichever approach is used, provided the discount rates are directly equivalent, the present worth values of the cash flows should be the same. What is the equivalence between the two methods? If the rate of inflation is 'a' per year, and the required rate of return is 'r' per year in actual cash flow terms, let the equivalent 'real' value of the discount rate be 'x' per annum. This equivalence is then given by the following:

$$(1 + a).(1 + x) = (1 + r)$$

This conversion can be used to convert or compare values on the different bases. However, as in practice inflation tends not to remain constant, direct conversions of cash flows to firstly 'real' terms and thence to their 'present worths' using the methods described earlier is more satisfactory for past and present expenditures and incomes and to estimate future cash flows directly in 'real' terms before converting to 'present worths' is probably the safest way to perform the assessments.

Chapter 8

Life Cycle Costing

Formats

Life cycle costing is a term that has been applied to a series of forms of
assessment of projects which, although in many ways similar to the
economic assessments described in Chapter 7, differ from it. The term has,
in the past, sometimes been applied to the full economic assessment
although in most forms it is less comprehensive. In its most common
formats it differs from the full economic assessment by omitting to take into
account the flow of income or benefits resulting from the project's performance.

At the time of publication there has been no agreed standard form for the
performance of Life Cycle Costing exercises. This is despite the strenuous
attempts by the International Electrotechnical Commission (IEC) via their
'Dependability' Technical Committee (TC/56) to agree such a standard.

Consequential costs

The difficulties arise, in part, from the plethora of forms of the technique
already in existence. Another major problem is whether or not the 'life cycle
cost' should or should not include the consequential costs which arise in the
event of failure of the asset or a component thereof. Such costs are
notoriously difficult to assess in advance. As a frequent purpose of such
LCC assessments is their use in selecting between tenders or in approving a
particular supplier's offer, predictive costs are inevitable at these times
when decisions have to be made by the user organization. That is to say, in

circumstances when the inclusion of these costs is particularly relevant. One possible method of establishing a common method of estimating such costs, to allow comparability between offers, would be for the user organization to dictate a formula for calculating the consequential costs from the assessed parameters such as reliability, maintainability or availability. The trouble with this 'solution' is that consequential costs can be so highly dependent upon the circumstances, especially in relation to the availability of other assets, at the time of the failure of the asset in this particular project.

For many projects the magnitude of the consequential costs are likely to be modest in comparison with the asset's capital cost and the direct costs of operation or maintenance. However, in a few cases the consequential costs can be a dominant factor. In a particular case in which the author was involved, a plant which had a capital cost of the order of £20 to 25M, had a potential for consequential costs of a very high order. During its expected 30 year life, the plant was required to be shut down routinely every third year for extensive maintenance and statutory inspections, each such outage lasting about 100 days. In the case of breakdowns of its major components, outages of even longer durations have been experienced. The consequential costs of each triennial routine outage were estimated as a cost on the user's organization of about £25M. That is to say, of the same order of magnitude as the original capital cost of the asset. Over a number of years the average additional shut downs for failures have amounted to about 60 days in each three year cycle with losses of proportionately similar or even higher magnitude. It is clear that in cases such as this the consequential costs are dominant to the point where they could swamp any LCC assessment. Their inclusion in such cases would depend heavily on the purpose for which the LCC exercise was being conducted.

The question of consequential costs is further complicated by the need for decisions on what should or should not be included under this heading. In the example given above we were discussing direct consequential losses. Such losses are the more obvious. However, there can be further sources of costs on the user organization which are indirectly attributable to the lack of 100 percent availability of the asset. These can arise from a need, of the user organization, to cover for downtime of the asset in order to meet its customers' demands for the product. Such cover may be provided from sources outside the project concerned. These may be provided by standby facilities, acquired separately or by retaining older plant which would, by now, have been retired had not this cover been necessary. Clearly both of these solutions will involve the user organization in costs not directly attributable to the project. Alternatively, should the cover be obtained by acquiring product from another source, the costs of such an exercise would be a more direct cost on the project and more readily apparent.

Standardization efforts

Although no internationally agreed standard on LCC has yet emerged, work, on producing such a standard, had been going on in IEC for several years. Various drafts have been produced but wide agreement has yet to be achieved. In several recent drafts an approach to establishing an LCC, involving breaking down all costs into its 'elements', has been suggested. Each 'element' is defined by its three basic parameters. These are (a) the sub-division responsible for the work, (b) the work involved, and (c) the life cycle phase in which the work is done. Effectively these deal with the 'who', the 'what' and the 'when'. Such an approach has the merit of ensuring that the whole project and thus the whole life cycle cost is covered. However, this is difficult to apply during the earlier phases of a project as the balances between alternatives, handled by different elements, will not have been established. Typical of such balances could be that between the use of automatic devices and the use of labour and directly operating features. Similarly the balances between predictive and breakdown maintenance and condition monitoring will not be known at an early stage. However, although every element may not be capable of prediction, estimates may be possible of the aggregated costs from a number of elements, e.g. the total maintenance cost of a particular component or assembly.

Buildings

The use of LCC in relation to buildings has been advocated within the building industry in the United Kingdom for several years. However, it has been noted on several occasions that its use, in ensuring that a building is completed to the optimum economic state for the user, is often ignored. This appears to arise from the peculiarities of the market for buildings. In industrial situations involving plant, only two parties are normally involved; these are the supplier and the purchaser/user. The user has then the ability, by means of his specification, and by his selection of a supplier, to obtain plant which gives him the best economic prospects. At the same time suppliers, conscious of the user's ability in this respect, are motivated to design and offer plant which meets the user's criteria. In the case of buildings however, there are commonly three parties involved. First there is a developer who has the building designed and built, usually to his

Table 8.1   A comparison of LCC and economic evaluations

Example

In this example inflation can be ignored and both costs and income can be expected to remain constant throughout the project. A commercial organization requires a 10 percent per annum rate of return on expenditures. The project is to provide and operate an asset for a period of 11 years (subsequent to the one year needed for the asset acquisition). It is due to be superseded by another project at the end of Year 11 of its operation. Two tenders for the asset have been received, A and B. In A the asset will cost £1M and the running costs will be £100 000 per annum. Tender B has a capital cost of £1 310 000 and running costs of £50 000 per year. The income from the project, from sale of the product, is calculated to be £289 300 per year, but will be received a year after that year's production. Just before placing an order to the asset, information is received that the replacement project may now be available one year earlier than previously expected, reducing the project duration by this amount. Examine the LCC implications of the two tenders in the two cases of 11 and 10 years operation and also their economic assessments.

In the following analyses costs and incomes are expressed in terms of £1000 shown as £k.

Tender A has costs of £k1000 in Year 0 and thereafter £k100 for years 1 to 11 in the original plan.

LCC = £k1000 + £k100 × 6.49506 = £k1649.506 (where the factor 6.49506 is the aggregated present worth factor over years 1 to 11).

If the project is reduced by a year:

LCC = £k1000 + £k100 × 6.14457 = £k1614.457 (where the reduced present worth factor is that for the aggregated years 1 to 10).

Tender B has an LCC of £k1310 + £k50 × 6.49506 = £k1634.753 for the full term of 11 years and

LCC = £k1310 + £k50 × 6.14457 = £k1617.223 for the reduced term.

The expected income over the original 11 years will be

Income = £k289.3 × 5.90499 = £k1708.2 (where the present worth factor now used is that for the aggregate of years 2 to 12).

If the production of the project is terminated one year early the income will be £k289.3 × 5.585969 = £k1616.0021.

Results

Original project of 11 years production

Tender A	LCC = £k1649.506,	Income = £k1708.2	Profit £k58.694
Tender B	LCC = £k1634.753,	Income = £k1708.2	Profit £k73.447

If project terminated one year early,

Tender A	LCC = £k1614.457,	Income = £k1616.021	Profit £k1.564
Tender B	LCC = £k1617.223,	Income = £k1616.021	Loss £k1.202

Commentary

It will be seen from the results that on the original 11 years basis Tender B has a lower LCC than Tender A but if the project is terminated a year early the reverse occurs. On a viability basis, at 11 years both tenders would be acceptable but on reducing the duration only Tender A has a positive return whilst Tender B results in a net loss. This illustrates both the care with which assessments should be made but also the benefits of using a sensitivity analysis.

specification and to his economic criteria of minimum cost within that specification. On completion, the building is then sold to a second party, usually a financial type of institution such as an insurance company or pension fund, who leases it to a third party, the ultimate user. In most cases the lease places the responsibility for both the running of services and maintenance on the occupier/user. With this contractual arrangement the first and second parties are only involved in the initial costs and are totally absolved from running and maintenance costs. They are therefore motivated towards a choice of designs and service systems which economize on these first costs even though these may not be the most economic overall. The user, who will have virtually no say in such choices, is thereby left to pay the extra costs of the less than optimum design and service systems. It is perhaps not surprising that many of the buildings for which great claims are made for energy efficiency and economic managements are buildings built directly for a user's needs and to his specification. Clearly this is comparable to that experienced in the acquisition of plant.

Table 8.1 gives an example which compares one form of LCC approach with that of a complete economic assessment method. It also shows the sensitivity to relatively minor changes in the planning assumptions.

Chapter 9

Suppliers

Users' views

It will be apparent from the foregoing chapters that the emphasis in considering the economic management of assets must essentially be from the point of view of the user organization. So how does the adoption of the philosophy by users affect and concern their suppliers? Firstly, it is important for suppliers to recognize that their customers, the users of their products, are increasingly becoming aware of their own needs to adopt this philosophy or, at the very least, many of its features.

The degree to which a particular user adopts the elements of economic management and considers costs and benefits over the entire life-time of the asset can vary very considerably. Furthermore, the extent to which the user makes this apparent to his potential suppliers, is also variable. At one extreme a large user organization with sophisticated acquisition policies and procedures may use formal enquiries and specifications which either call for such details to allow it to perform economic appraisals or simply to require offers to be accompanied by life cycle costing analyses. In the latter case, rules for conducting the LCC analyses should be clearly laid down by the user. At the other end of the spectrum, a motorist seeking a new car may be equally interested in fuel consumptions and servicing requirements as in the car's features and price. The weight he gives to these various aspects will seldom be expressed to the salesman in any clear form. However, a good salesman should be fully aware that the potential customer may be viewing his offerings in this light.

Suppliers' reputation

The reputation of a supplier will depend to a large extent on how well he has taken the user's needs into account in the design and production of his products. As customers become more sophisticated and start viewing their assets from a life-time economics point of view, a supplier's reputation will become even more susceptible to his attitude and performance in this regard. It will also behove a supplier to demonstrate to his potential customers the extent to which he has taken their needs into account. In recent years much has been done to improve the quality of products. Demonstration of the supplier's intent in this regard, by obtaining certification of conformity with ISO 9000, can be useful in persuading customers that the supplier is taking the question of quality seriously. That the supplier is taking other requirements, needs or aspirations of the user into consideration may require tackling in different ways. These may take the form of LCC exercises, reliability and maintainability studies or simply explanations of why particular features had been chosen and a comparison of these with alternatives which had been fully considered and assessed inferior. These may or may not have been requested by the potential customer. In cases where such features are deemed to be critical for a user, these should undoubtedly have been requested at the same time as his formal enquiry. The most appropriate format and vehicle for conveying to prospective customers details of such efforts on behalf of their interests, will depend upon their circumstances. Suppliers should make sure that both the exercises and the transmission of the results to prospective customers are undertaken in constructive and effective ways.

Application to supplier's own asset

Quite apart from the above, the supplier should also seek advantages from ensuring that the cost of producing his product is as low as possible, consistent with it achieving the performance required in every respect. Such action should be to his advantage in two respects. Firstly, it gives the supplier the prospect of increasing his profit and secondly, by reducing the cost of the product, he can expect to increase his share of the market for the product. At the same time the customer should also benefit from the reduced initial cost and also from reduced replacement costs.

Many techniques for improving the production process have been promulgated in recent years. Many of these have been derived from Japanese practices such as JIT (just in time). Applied in appropriate circumstances, these should contribute to minimizing the cost of production. However, action should also be taken at the earlier design and development stages. This could include the use of 'value management' and 'functional analysis'. These two techniques are currently being considered by a European Standards Technical Committee with a view to publishing a guide or guides on these techniques.

Communications between suppliers and users

In addition to the procedures and techniques described above, there is one important task that the suppliers should undertake. Indeed this can be vital to ensuring that certain procedures and techniques are performed most effectively. This task is to ensure that communications between the supplier and the users are efficient and effective in conveying all the data, information and views between the two parties. Whilst this is particularly true when a user is seeking to acquire a highly specialized asset or component, it still remains true when a supplier/manufacturer is offering a product to a mass market.

In the case of a specialized product being offered to a single specialized customer, this communication exchange should not be limited to the earliest contacts and the formal specification and enquiry. Co-operation in the design and development stages can prevent the pursuit of certain optional courses which are, in the long run, likely to be abortive. In this connection the supplier should take advantage of the fact that the user organization may well have a wider, or at least different experience of the type of asset being considered than the supplier and his staff. Co-operation between the supplier and user will not only avoid some of the effort being wasted on abortive work, but may reduce the overall time required to complete the asset or components being supplied. This is clearly in both parties' interest.

Market research

The case of a product developed by a supplier and offered to a wider market is different. Clearly a one to one consultation process with each and every potential customer is not a feasible proposition. Nevertheless, consultation with users is equally desirable, although other types of communication will have to be employed. The general name for such investigations into the views of the users is market research. As conducted in the majority of cases these days, the feedback tends to be very limited and consequently of limited value to the designers. This arises because there is a tendency for very generalized questions to be put to a very large number of customers or potential customers. Relating the answers to the individual circumstances may not be possible or at least very difficult to achieve. Because questions are generalized, replies seldom give a full understanding of the users' views. There would appear to be a strong case for designers to be directly involved in the market research being carried out on their product or proposed product. This should enable questions to be phrased in a manner which elicits responses which will directly help the design team. Directing the questions to a more selective sample of customers, those with the potential to give reasoned and useful answers, rather than a random sample of users, should be more effective in this regard. It is known that one UK automobile manufacturer has adopted such an approach by holding special previews of their proposed products to members of an institution of qualified engineers. Overall, to ensure that market research does achieve a feedback that conveys the users' viewpoints in a way that is helpful to the supplier organization, great care should be taken in the design and arrangements for the conduct of this research. If a manufacturer is really needing to know what type of ignition system is preferable, it is no use devising a questionnaire aimed at finding out preferences in body colours.

Chapter 10

Contributors and Involvement

Project initiation

A project, requiring the acquisition of a new asset, is invariable initiated in a user organization. As the concept is developed in the user organization, the detailed requirements of the asset are developed to meet the various needs of the organization over the life-time of the project and the asset. In the first instance the product and hence the basic functional requirements of the asset have to be established. As far as the product is concerned, the product desired at any future time must be determined, if necessary by means of market research. This requirement should be used to determine the capacity required of the asset and its performance in terms of availability etc. Relying on these basic details as the sole basis for seeking to acquire the asset is unlikely to ensure that the asset will prove either truly effective in meeting the overall objectives, or anything close to an economic optimum.

To achieve these latter objectives, further requirements will be identified as necessary or desirable. To ensure that these necessary or desirable further requirements are met, the basic asset specification will have to be supplemented. However, in the case of these extra desirable requirements, a balance must be drawn between the cost of meeting them and the value of the performance enhancement that they will achieve. This latter has to be evaluated in terms of the resultant effects on both life-time costs and any income accruing from the sale of the product.

Sponsorship

Whilst one person will normally be acting as the original sponsor of the proposed project, the input of these refining requirements should normally come from other specialisms within the user organization. This, at least, will be the situation within a large user organization, contemplating a major new project. For lesser projects, the considerations are the same, but fewer people may be involved and the degree of consultation in formulating the specification requirements will be correspondingly reduced.

In the initial stages of the conception of the new project, sponsorship may lie in any one of a number of sections within the user organization. At any one time a number of projects may be undergoing development. These will, however, all have a common feature of being likely to require investment in a new asset. This implies the expenditure of or commitment from the user organization's financial resources. Inevitably this renders all the various new projects being devised for the organization as being in competition with one another for the limited finance that the organization can make available.

Authorization

For the organization's senior management to be able to determine which new projects should be authorized to proceed, all the competing concepts must be arranged as formal proposals in a common form identifying the extent and timing of the financing required and, on a common basis, the likely economical advantages to the organization.

In the first instance the concept must be fleshed out in more detail and, in particular, for all costs and the likely incomes or savings from the proposed project to be identified. This will involve both the initial and running costs of the new asset, the asset's performance and the likely output of product it will achieve. Where these aspects are critical to the project's overall economic performance they must be evaluated realistically and using the best possible data. In particular underestimates of costs or overoptimistic estimates of the value of the product output should be avoided. A somewhat pessimistic approach to the assessments is usually prudent.

Individual contributions from the various functional departments to the initial evaluation exercise and to all subsequent exercises must be made on a common basis especially regarding financial inputs and discount rates. Other features such as asset/project life-time should be part of the common basis both for inputs to the economic assessments and the judgements to be made on alternative courses of action.

The next stage is for the proposal to be submitted in formal form for the organization's approval to proceed. This proposal should give the full details of the financing requirements throughout the project life. A statement by the organization's finance specialists should be included in the submission together with their observations on the availability and sourcing of finance and their recommendations regarding an appropriate discount rate to be used in the evaluation of its validity. Clearly this requires consultation between sponsor and finance specialists in advance of the formal submission stage. It should also give the economic appraisal of the project as a whole. To allow the competing proposals to be compared, the basis of this economic appraisal may well be required in a standardized form and using a standard prescribed discount factor. It is helpful if the results of a sensitivity analysis of the factors used in the economic analysis are included, as the standardized discount rate used for comparison of proposals may not be fully appropriate to the circumstances of that particular proposal and its manner of financing. The inclusion in the submission of statements of support from other departments likely to be affected by the project is usually found to be helpful.

Acquisition

If the project, after review and comparison with alternatives, is given approval by senior management, the acquisition stage for the asset can commence. Together with their approval to proceed, senior management should select and appoint a 'Project Manager' for the 'acquisition phase' of the overall project. Common usage of this term 'Project Manager' is in some ways unfortunate as it tends to concentrate the term 'project' upon this 'acquisition phase'. In terms of proper economic management, it must be remembered that the 'project' does extend through the 'useful life' and 'disposal phases'. Economic evaluations and the selection of alternatives must take this into account. Far too often this is forgotten or ignored and optimization of costs arising in the acquisition phase are the only ones considered by the Project Manager and his team. This should be avoided by the application of an effective form of monitoring of their performance and decisions, and ensuring that the operations and maintenance departments are encouraged to become involved. Their contributions, especially regarding the compatibility of the new asset with existing assets, and their fundamental policies can be very valuable. The first practical task of the project team is to prepare the specification or specifications for the complete asset. These must include the principal functional requirements as identified in the scheme approval. There must also be included any additional requirements identified by specialists during the 'concept phase' and included in the economic assessment submitted for the project approval.

As the responsibility for the project may well have changed from the sponsor, in the 'concept phase', to the Project Manager in this 'acquisition phase', the consultations with and inputs from specialist staff, may well have to be repeated. Furthermore, these requirements will have to be spelt out in fuller detail and prepared in a form suitable for inclusion in the specifications. Inputs from both the Production and Maintenance specialists will of course be critical in this task. Their inputs will need to include statements of the user's established policies where these could affect the detailed designs offered by potential suppliers. For example, choices between automatic and manual controls of various devices might have been established as a result of the user's prior experience. Similarly, the maintenance policy and the extent and nature of external maintenance resources must be revealed. These sorts of information will be essential for potential suppliers to be able to offer optimized designs of the asset. Another essential input into the specification will be the user's expectations of the utilization of the asset through its required life-time.

The specification should also include, where relevant, requirements based upon the distilled wisdom of experience as given in various standards. The most useful source is likely to be any appropriate national standards. International and company standards can also be relevant and suitable for inclusion. European standards issued by CEN, CENELEC or ETSI may establish design features which ensure that the asset or component conform to the terms of European Directives: so for the appropriate European countries these should be taken into account. As far as the United Kingdom is concerned, these European standards are, in almost every case, covered by an identical British Standard mostly in the 'BS EN XXXXX' series.

Potential suppliers

Concurrent with the preparation of the specification(s), the project team should establish potential suppliers for the asset or part thereof. In the light of European Union Directives, if the user is a public body, this may involve a search for potential suppliers from other EU countries. Dependent upon the nature of any potential supply contract, a degree of vendor assessment may be desirable before including a particular supplier in the list of organizations to whom the specifications should be sent with a request to tender. This vendor assessment should include such features as technical ability and experience, reliability and financial stability.

Before issuing the specifications together with their required extent of supply, the existence of available components for the asset already held by the user organization should be checked. If any such components are found, the impact of their diversion to this new project should be determined, especially the effect on the organization's overall economic situation. Only if this is advantageous, should the component transfer be authorized and that component withdrawn from the specification and the request for tenders.

With the issue of the specification and the accompanying request to tender to potential suppliers, the principal activity shifts to the supplier organization. Its immediate task is to provide a tender indicating the principal features of its proposed design, (and usually its alleged merits), the extent of proposed supply and the price. During this period, the project team will be required to answer queries by suppliers to a variety of problems posed in varying degrees of formality. These can range from interpretations of the specification itself, to seeking guidance on the user's views on specific design options. Questions of the latter type may well need to be referred to the user's specialists before a reply can be framed.

Tender assessments

The next major involvement by the user organization arises when tenders have been received from the potential suppliers. In the first instance the conformity of the tenders with the specification must be checked and variations identified. This will be the principal task of the Project Manager's team. In addition, the tenders should be scrutinized by the various specialists in the organization, especially those who had been consulted and asked to contribute to the specification. These specialists may well find areas in the tender which give rise to doubts in their minds either as to what precisely is being offered or the implications of varying the offer to accommodate a preferred feature in the design or a proposed method of working. The likely cost/savings of introducing these particular features and the probable costs of amending the tenders to conform entirely with the specification should be estimated either by the project team or specialists. They should also estimate any additional life-time costs, or alternatively savings, arising from any other aspects of the design. These estimates, together with the tendered prices can then be used to compare tenders in a preliminary tender assessment. This may eliminate some of the offers from consideration. The suppliers remaining may now be sent a questionnaire to clarify any points of doubt about their tender and to submit prices for either correcting their tender to conform with the specification or to incorporate additional or special features now deemed desirable. The replies to the questionnaires should now enable the project team to modify their preliminary tender assessment using corrected cost figures and establish the preferred supplier with whom to negotiate a formal contract for the asset or component concerned.

Letting of contracts

Before any contract to a preferred supplier is actually let and the user organization committed to expenditure, a final economic assessment of the project as a whole should be carried out using the costs and data to be agreed in the contract. Only if the results of this assessment remain satisfactory in terms of viability, and any other criteria set when the project was approved, should the project be allowed to proceed. This is a key point in the overall process.

Environmental aspects

For any new project, especially an industrial project, the impact of environmental risks must be taken into account in the economic assessments. This includes risks from the asset itself as well as from its operation and disposal. Many environmental risks and the laws and regulations to eliminate or limit these are now well established and the impact of these should be taken into account at the concept stage. Further risks and the likelihood of restrictive legislation or regulations can normally be anticipated for some time in advance.

For the already established risks and regulations, suitable clauses should be incorporated in the specification. If the project is of a long term duration, it will also be prudent to include requirements for the avoidance or amelioration of the further environmental risks which might be the subject of future legislation or regulation within the planned life of the project.

Documentation

Alongside the development of the asset and its manufacture and supply, documentation must be prepared and submitted. The requirements for this documentation must be included in the specification for the asset. Alternatively, if a separate contract for some of the documentation, such as the technical manual, is proposed, the asset specification should contain a

Table 10.1 Normal responsibilities for project in user organization

Phase	Responsibilities for economic management
Concept	Sponsor
Acquisition	Nominated Project Manager
Useful life	Nominated Project Manager until handover to Operational Manager for Asset. This is usually the manager of the site on which the asset stands.
Disposal	Operational Manager until handover to specially appointed Disposal Manager or purchaser of asset/ component if appropriate

requirement on the supplier to make information available to the contracted authors. It is most important that submitted documentation is adequate for the needs of the user organization. As far as the Technical Manuals are concerned these should meet all the user's requirements. This will include all the sections appropriate to the particular project as identified in BS 4884, Part 1 and ideally prepared as advocated in the guidance standards BS 4884, Parts 2 and 3. To ensure that the Manual or Manuals are suitable for the user, it is usually desirable that drafts are submitted for consideration and checking ahead of the commissioning stage of the asset.

Erection or installation of the asset may or may not be included in the supply contract or may be undertaken by the user himself. In either case it is important for the Technical Manual to include the necessary instructions. This is necessary because, even if the supplier erects/installs in the first instance, circumstances can often arise during the 'useful life phase' and when the supplier has long left the site, when the user needs to re-erect the asset or major parts thereof.

Commissioning tends to be a joint activity between the supplier and user. For this reason it is important that the commissioning procedures are jointly established and programmed and, ideally, agreed in written form. For the supplier this stage is that in which he proves the performance of the asset and its components meets the requirements of the specification and any offer made in his tender. For the user's production personnel it is usually their first chance to operate the asset and as such is an important step in their familiarizing and training.

As the project proceeds, the responsibilities for it will change generally as shown in Table 10.1.

After the asset has been fully commissioned, the 'useful life phase' commences. This is often known as the 'operation and maintenance phase'. By this time the major expenditure on the asset's acquisition will have been spent and thus the user organization will have been irrevocably committed to a significant part of the project's life cycle cost. As a result there is normally little point in carrying out, at least in the early days, further detailed evaluations of the viability of the project as a whole. Accordingly in this phase most evaluations will be of a comparative nature in which the effects that alternative courses of action will have on the user's economics are assessed. At any time this could include a comparison between continuing the project and closing it down prematurely.

More frequently, the comparative evaluations will be to determine which of a number of alternative courses of action is economically preferable to the user organization. These can include choosing between maintenance policies and procedures and determining spares policies and stock holdings. Determination of the 'best' operational plan for maintenance and production should also be based upon the overall economics of the project. To achieve this it is important that the department carrying out this planning is given the necessary information regarding the true effective costs of production and the marketable value of the product at any given time. This will inevitably call for inputs from the production function but also those responsible for both marketing the product and the purchase of raw materials for the process. When production/operation initially commences, the data used in these evaluations will still be based upon the original estimates, many of which will be merely 'best guesses'. As experience grows with the asset, these estimates should be up-dated in the light of actual performance. It is therefore important that the data, required to enable the inputs to these economic comparisons to be accurate, be initially identified and then collected and recorded. Far too often data collection is limited to that required to feed senior management with general indications of overall performance or to impress them. Concentration on the data needed to effect the best possible evaluations and thereby the best managerial judgements on the part of those directly involved in the operation, maintenance, servicing or planning these activities, is likely to be more profitable. Clearly all these sections of the user organization must be involved together with all the other sections mentioned above.

Data handling

The amount of data involved, its analysis and keeping it up to date can be an expensive exercise and may warrant the setting up of a special records section. As with all other activities, the extent of and methods used in data collection and recording should be critically examined for justification in the circumstances of that project. This is difficult as there is often no measurable direct return. Because of this and in the light of the expense involved, such a section is vulnerable to calls, especially from those not using the data, for such sections to be closed down or even from being set up in the first place. However, the data are essential for the accurate evaluations on which proper management of the asset will depend. It will be seen that virtually all sections of the user organization should contribute to this pool of data. In turn, this must be made available as required by any section. For any one section to consider his own input of data to be 'owned' by that section and released only by itself, is unacceptable. Whilst this is seldom found in and between operation, maintenance and planning sections, it is commonly experienced where some other functional sections are concerned. Senior management should ensure that this does not occur both by the way they set up the records system and by monitoring that the correct data flows and priorities are maintained.

End of life

As the asset or project approaches the end of its planned life, a series of evaluations should be carried out into the alternative courses of action for the project as a whole. These will include;

(a) Continuing the project to the planned terminal date.
(b) Continuing beyond the original terminal date.
(c) Closing down the project prematurely with or without a replacement project.

Such evaluations will require inputs from virtually all sections of the user organization. If a replacement project is included in the final option, this must be evaluated in the same way as an original concept and must be based on both up-dated technology and performances as well as up-dated costs. All sections will have to contribute to this both from a technical and costs point of view.

It is not uncommon for the asset or a major part thereof to suffer a major mishap during the course of its planned life. In such circumstances, it would be prudent to carry out similar economic evaluations of the future courses of action, additionally taking into account the probable costs of restoration and the time it will take as well as any losses in income. Clearly these should precede any commitment to any restoration work. The costs and times to restore the asset may themselves have a number of alternatives determined by the desired extent of the restoration.

Modifications

Modification to the asset or its parts may be proposed at any time during the useful life phase. These may emanate from the suppliers directly or as a result of user's complaints of inadequate performance. Viewed qualitatively, such proposals invariably appear to be desirable, as they are claimed to improve performance where it is deficient or suspect. However, before adopting such proposals, a user organization should first evaluate what the economic implications to itself will be. In many cases the cost of the modification can be relatively modest but the downtime needed to carry out the modification could be very high and the loss of income significant.

In one case in the author's experience, not only was the above true but the organization would have suffered significant further costs, not apparent when only that asset and its management was considered. These costs arose from the asset being rendered different from otherwise identical assets in other managerial units within the same organization. As a result, the modified asset would be unable to use the jointly held pool of strategic spares which had been called upon at frequent intervals in a tactical role. As a result future outages of the asset, when these spare components might have been used, would inevitably be prolonged. In this case the modification could not be justified for the whole of the family of nominally identical asset units. Although apparently justified for the one management unit, when viewed independently, the adoption of the modification at this one site led to an overall 'loss' to the organization as a whole. This example is a good illustration of how essential it is for evaluations to be carried out from the organization's overall point of view and for all parts possibly affected to contribute to the economic evaluation.

Disposal

Planning for the 'disposal phase' should start before the end of the 'useful life phase' especially when special problems, such as radioactivity, could be involved. After the asset has been completely de-commissioned, re-usable materials should be recovered and either diverted to a new use or stored. This will apply to both feed materials and parts of the asset. Preference should normally be given to items required by or could be used by other sections of the user organization. Other parts not usable by the organization may still have useful life in them and might be saleable to other organizations. Ideally the recovery exercise should be completed before demolition and scrapping of unwanted materials takes place. In any case, these latter activities should not proceed in a manner which jeopardizes useful material and its re-use.

Both the processes of dismantling the asset and disposing of it can give rise to environmental hazards. Some of these could be the subject of restrictive legislation or regulations, many of which may have been introduced since the asset was designed and built. Meeting these regulations may limit the optimal courses of action for carrying out the disposal processes. Specialist knowledge may be required and generally it is better and cheaper in the long run to seek advice from the regulatory authorities. As in other phases, a choice should be made between acceptable alternative options on the basis of economic advantages. This will normally principally concern the comparison of costs because production and therefore income to the project will have essentially ceased. Nevertheless, returns from the sale of surplus materials and scrap must not be ignored. Neither must the effects of time on cash flows be disregarded. For example, the completion of the disposal phase and the ability to sell or rent a site could have a financial value to the user organization.

Hitherto this chapter has illustrated the participation of the many functional sections of the user organization in providing inputs both technical and of estimated financial values to the decision processes in the project's management. These decisions are invariably interrelated and thus there is a need for all parties to re-assess their contributions, both technical and economic as the project develops. Clearly this will require very extensive communications between the parties. This will require a system of efficient communications, efficient both in terms of physical equipment and in established procedures. The effectiveness of the economic management process depends on these efficiencies. The importance of this is such that it should be recognized, set up and maintained by senior management. The willingness of the parties to communicate and overcome the tendency to be possessive with information is equally important and this may require a cultural change on the part of some employees. The establishment of suitable training and enforcement may be required. This again will require senior management's support to ensure that it is achieved.

Specialists not required

Unlike certain management concepts, terotechnology or economic management of assets does not require the introduction of specialist staff into the user organization. Each functional department in the organization must contribute their technical expertise to the management of the project. However, in addition they are required to consider their contributions in the light of their overall effect on economics of the project. This can only be achieved if each functional department develops an appreciation of the contributions made by the other departments especially in relation to their effect on the overall economics of the project. These are merely extensions of their existing technical inputs and do not call for specialist knowledge. It may, however, call for a cultural shift in the way in which all the departments of the user organization need to approach their work. As such senior management should not merely support the concept of economic management; they should insist on its adoption and ensure procedures maintain it.

Communications are not limited to within the user organization. For example during the 'acquisition phase' there will be extensive communications between the user and the supplier(s). Again these should be as free and as extensive as possible within the constraints of the commercial relationship. This invariably affects financial data rather than technical aspects. However, offers of alternatives from the supplier should always be accompanied by cost implications as far as the supplier sees them. In turn the offers should be judged by the user in the light of their overall effect over the project life-time as well as on their technical merit. Certain sensitive cost information might have to be withheld from the interchange and only the resultant decision transmitted.

Chapter 11

Effect on Working Practices

The adoption of economic management of assets should not have a major impact on the working practices of a user organization. Nor should it have any direct influence for changes in the organizational arrangements. However, the adoption of the principles of this management technique may show up weaknesses both in the organizational structure and in the working methods. These should, of course, be dealt with in their own right.

Chapter 10 illustrated that virtually every functional part of the user organization has a part to play in the correct economic management of a project and the assets it employs. It also indicated the manner of that involvement. This chapter expands on the likely changes that this will require in working practices.

Table 11.1 summarizes the principal contributors from the various management functions in the user organization.

Communications

The first and probably most important aspect involves the communications between the different functional departments of the organization. In this, the whole of the organization is affected. Firstly data must be made readily available to all other departments who need it either for their direct use or to form a basis on which they can formulate their own views on any matter on which they should be contributing their expertise. For this to be effective, there may have to be much greater freedom in the exchange of information between functional departments than has been experienced in the past. This could include the acceptance of less formal communications

Table 11.1   Principal contributions to economic management

---

(1) *Higher management*
- Final authority for approval of project and contract release.
- Insistence on proposals being supported by economic assessments.
- Promulgation of financial criteria for project proposal assessments.
- Promotion of communications between departments with minimum restraints.

(2) *Finance department*
- Recommendations to higher management on common financial criteria.
- Recommendations to sponsors and higher management on financing proposed projects both on sources and rates of return required.
- Check for higher management that economic analyses are correctly assessed.

(3) *Production department*
- Advise on manufacturability of product.
- Advise sponsor and Project Manager on aspects of design of asset and manning.
- Develop production plan in conjunction with maintenance department and output plan required by sales/marketing departments.
- Establish operational records for asset and components.

(4) *Maintenance department*
- Advise sponsor and Project Manager on aspects of design of asset and maintenance staffing and production department on maintenance outage requirements.
- Develop maintenance policies and plans in conjunction with production department.
- Advise on strategic spares requirements and initiate minor projects for those not being purchased under the initial contracts.
- Establish routine spares holdings and restocking procedures.
- Establish history records for the asset and its major components.

Note: These contributions are given in more detail in BS 3843, Guide to terotechnology; Part 3, Guide to the available techniques.

between departments and a lowering of the levels of authority needed to request and receive the data, etc., on the one hand, and to release the information from the party holding it. This will arise simply because of the increased quantity of information flow on which proper economic management depends. The time needed to carry out such exchanges of information formally would inevitably be too long and the information would be received too late to be of value. Formal methods would also inhibit an iterative approach to the contributions made by the department receiving the information. Furthermore, formal exchanges may well prove unnecessarily costly in terms of time and effort.

Access to data

Clearly, as stated in the last chapter, this development in a normal working practice, will both require, and help to establish, the principle that data and information does not 'belong' to a particular department collecting and holding it but should be readily available to anyone in the organization who needs it.

It is appreciated that the above 'freedom' militates against the frequently held, but not always justified, claim for security of information. However, in essence, it does conform to the principle of 'need to know'. The conflict between the principles of freedom of access to information and security, can be eased by a proper identification of what data or information are really critical to the organization's interests. Far too often, instead of properly assessing whether an element of data or information needs to be secure and the degree of security needed, a blanket and, much higher than necessary, classification of security is imposed on all information. Secondly, the 'need to know' aspect can be prejudged to a large extent by consultation between the various departments and therefrom establishing 'standing' arrangements. The problem is, of course, eased if the amount of classified information is itself reduced to the essential minimum. It will also help if departments were to make distinctions between the data and information that they need routinely and that which they could need when a specific topic arises. This latter category could be the greater amount of data and information, by far. An example of this latter category can be taken from the author's own experience. At one time, when investigating a proposed project to hold a strategic spare HP turbine rotor for a family of large machines, a knowledge of the past history of such components from similar machines was needed. Analysed data from this history were to be used as an input to the economic analysis for the project. Having completed that exercise, the particular data were not wanted again until a fresh review of spares holdings was initiated several years later. The problem with occasionally required information such as this, is that, because it is not required routinely or even frequently, assumptions are made that it is not worth collecting and recording it. However, when it is needed, it is essential for the proper assessment and management of the project. Its absence, when needed, could give rise to either wasted expenditure on an unjustified project, or missing an opportunity to invest in the project which, in the long run, would achieve considerable operational savings.

Whilst the actions recommended above do not totally resolve the conflict between access and security principles, the position can be ameliorated further if the general climate of openness between functional departments were to lead to an improved appreciation of the individual departments' contributions to the overall economic management processes and thus their needs for data and information.

Data handling

It will be seen from the last paragraph that the amount of data that may be needed for feeding into the economic analyses of an organization's projects can be very extensive. As economic management develops in an organization so the flows of data will substantially increase. It therefore becomes an economic need for the organization to increase its efficiency in collecting, recording, analysing and co-ordinating, retrieving and communicating these data. When such operations were carried out manually, the cost of carrying out these functions, to the extent now needed, was difficult to justify and was often deemed uneconomic. However, modern technology has rendered most of these operations capable of being carried out automatically using computers etc., and is no longer manpower intensive nor expensive. This completely changes the economics of these operations. The automation of the data handling operations will, of course, require a change, generally a simplification, of the tasks of those carrying them out. In turn this can require changes to the non-data handling tasks that such personnel may be required to perform. In some cases this may relieve personnel of some of their work-load and leave them free to carry out additional tasks. On the other hand, it may be necessary to carry out a substantial reorganization of their tasks and/or modification of their equipment.

Operational data collection

An example of the last type of modification to working practices arises in the case of operators of machinery and processes. In the past, operators were usually called upon to visit all parts of the machinery or process to read instruments and record their readings on log-sheets. Now the readings can be taken automatically and fed directly into a data processing unit. The merit of the original system was that personnel had to visit all parts of the unit and would thus see any developments of mishaps or distress in the machinery or process or any hazard arising, such as a fire or the risk of a fire from leaks or rubbish. This allowed urgent action to be taken. With the automation of readings, the need for operators to visit the whole plant routinely is removed. However, alternative tasks or equipment must be introduced to ensure that those circumstances formerly detected visually do not go undetected.

Centralized data handling

With the larger amounts of data and information to be handled and the equipment for carrying this out, it is possible, in some cases, that a centralized data handling unit, serving the whole of the organization, could have advantages. These may be, in terms of economies of scale, for the data handling and storage equipment, as well as in manpower, especially as regards Information Technology specialists. The adoption of such an approach will inevitably modify the working practices of the data contributing departments at least. This centralized approach will demand, as a minimum, a common approach, throughout the organization, of the manner of storing data and information. In turn this will help in the training of recipient sections. These may well be supplied with computer terminals from which they can directly access the data and information they need, without recourse to written or verbal requests. Clearly the above ideal situation would require very careful and detailed planning and programming to ensure that all departments' needs can be met by the IT system purchased by the organization. Experience shows that this is not always easy or cheap to achieve.

Data analysis

The analysis of data and information for a particular purpose may need to be different to its analysis for other purposes. For this reason there can be advantages in arranging for the storage of most data and information in a relatively 'raw' form. Analysis can then be carried out by the recipient/user department in a manner that suits its particular purpose.

Using, as an example, the HP turbine rotor spare scheme mentioned earlier in this chapter, the historical data of failures of and maintenance to such components had to be examined in detail. This was necessary as only those incidents in which a spare could be usefully deployed to save downtime were relevant to a case for a spare. This, of course, was very different from the overall statistics of rotor failures and preventive maintenance exercises and called for a judgement to be exercised on each and every recorded outage. These judgements had to be based upon extensive experience of the maintenance of such components. This analysis was very different from those carried out for reliability statistics of such components or for the general indication statistics provided for management based upon the total loss of availability due to such components. Neither of these analyses would provide a suitable input into an economic justification for a spare component. Indeed most economic analysis exercises will call for similar discrimination of the recorded base data.

Should this analysis of data become predominantly the responsibility of the recipient rather than the originating or holding departments, there may be a significant shift in both work-load and data processing tools between the departments which could further affect the working arrangements and staffing of both.

Other techniques

The increases in the extent of communications and data handling and processing are the two principal causes which have a direct impact on working practices as a result of adopting this economic management of the assets. However, the philosophy of economic management also calls for the consideration and implementation of all other techniques which can contribute to minimizing overall costs or maximizing the organization's overall profits. Most of these techniques have been developed and promoted as separate philosophies and practices in recent years. Many books and papers have been written about these and therefore they will not be discussed in detail in this book. Unfortunately, some of these techniques have been promoted in such a way as to suggest that they would be a sure way of optimizing the organization's activities and make no reference to limits in their application. However, to ensure that the economic optimum for a project is achieved elements of all these techniques may need to be adopted. That is to say adopted to such an extent that, when in combination with all the other techniques, the economic optimum is achieved. This may well require only partial adoption of any individual technique and that this should not be taken to its extreme. For example, the technique known as JIT, when taken to extreme, results in the elimination of storage both throughout the process and also at the start and end of it. However, the provision of some storage, especially at the end of the process, could overcome variations in the demand of the product without calling for the asset to be rated high enough to be capable of meeting peak demands. This might therefore lead to reduced acquisition costs whilst avoiding losses of sales of product as a result of being 'stock-out'.

Chapter 12

The Nature of Projects

Commercial projects

Throughout this book we have used the general terms 'project', 'asset' and 'product'. In most cases these terms and the relationship between them will be well understood, especially where a commercial venture is concerned. For example, an organization decides to manufacture and sell washing machines. The 'project' is to produce the 'product', washing machines, and to sell these at a profit whilst the 'asset' is the plant needed to fabricate the washing machines from the 'feed' or 'raw' materials. This reference to 'raw' materials will include such 'bought out' components as the electric motors and the drive belts which have been manufactured by sub-contractors. As described in an earlier chapter, the 'asset' will include not only the fabricating and assembly plant but also such necessary items as the buildings necessary to house the plant and the processes and those needed to store the raw materials and the finished products; that is to say the washing machines.

Service projects

In other cases, such as a services project, the position may not be so clear. Let us consider a case where a government wishes to establish, within its air force, a squadron of new fighter aircraft. Within such a project there will normally be two separate 'user' organizations. The first and initiating 'user' will be the air force requiring the new aircraft. This organization's 'project' will be to acquire, deploy and operate the aircraft needed to fulfil a particular segment of its defence commitments, the 'product'. Its assets are more difficult to identify. However, they would comprise the offices etc., needed to specify the requirements for the new squadron in terms of numbers and performances, to handle the contracts with the suppliers and, in material terms, all those necessary parts to operate and maintain the squadron. That would include the landing fields, runways, hangars, control towers and the aircraft operating and maintenance facilities. The 'sub-contracted' aircraft from the supplier will be, in effect, the 'raw' or 'feed' material in this project.

In this case the other 'user' involved will be the aircraft manufacturer. His 'product' will be the aircraft themselves and his 'project' will be to produce them to the specification given to him. His 'asset' will be the factory needed to manufacture the aircraft and all the necessary manufacturing plant it contains. Associated facilities such as a drawing office or test facilities would also be included to the extent they were needed for this project.

Other service type projects will be related to the 'product' being a service which may or may not be provided after payment in some form. Included in this group would be 'maintenance services', in various forms, such as for burglar alarm systems or domestic appliances. It would also include governmental services such as walfare provisions. In this latter case the 'product' would be the assessment and disbursement of the welfare payments, and the 'project' will be the carrying out of the government's welfare policy. The assets would be the office buildings used for this task together with the equipment used to carry it out, such as desks and computers. The 'feed' materials for this project would be the data inputs to the system.

Modification projects

Finally, a project may be to modify an existing asset. This will normally be done to improve the output or some other characteristic of the asset such as its efficiency or maintainability. For this type of project the economic assessments should be carried out in at least two ways. Firstly, the project should be assessed as a stand alone prospect. In this, the costs and savings should be assessed in terms of the differences between their values after the modifications have been carried out and their values, as now foreseen, if the modifications are not carried out. This assessment will establish if the modification is worth while in its own right. As a further check on the proposal a re-assessment should be made of the original project for which the purchase of the original asset was authorized. Should the original assessment have been overoptimistic, and experience to date shows it was not truly justified, fitting the modification may just bring the overall scheme, with the modification, into viability. However, if the viability of the overall project is so marginal, these two assessments might be compared with assessments of further proposals such as to wind up the original project, immediately or at some future date, without carrying out the modification and initiating a new replacement project. This new project will, of course, have the advantage of being able to use the latest technology and will have better data bases upon which to estimate costs and performances. When carrying out these assessments, it is essential that the overall economics of the user organization are taken into account. This may require consideration of other proposals incorporating elements of a number of the above proposals to achieve the optimum for the organization.

Continuation and extension projects

The reasons for initiating a new project are both many and highly variable. When an organization has a product for which there is a continuing demand, such as say table salt, the original project will invariably come to an end of its life before the demand for the product ceases. In such a case, a replacement project will be needed to take over the supply of the product. This will normally occur when the original asset reaches the end of its economically sustainable life.

Replacement projects

In other commercial enterprises, the demand for the product may disappear completely. This, rather than the wearing out of the original asset, may be the determining factor that limits the life of the project. In such a case the organization will need to identify a new project with a completely new product in order to stay in business. In these cases, elements of the original asset may be usable in the new project, possibly after a degree of refurbishment. Associated buildings can often be re-used in this way. Normally, because of the organization's experience and strengths in certain core activities, new products will tend to be aimed at similar markets to those to which the original product was supplied. For example, if the original product was a toilet requisite, the new product will probably be another toilet product, to be sold to the same market and thus to be able to use the same marketing organization. In other words, as well as using elements of the original asset, continuing use of other elements of the organization may well be possible.

Expansion projects

The above examples are effectively projects in which the new product is a replacement for the original project's product, thus maintaining the organization's level of profitable activity. However, the organization may wish to expand its activities or the level of an existing activity. This will normally require the initiation of an additional project. In such cases, much of the above will equally apply, except for the termination of the original project. Additionally the new project must be evaluated in its own right and, as far as any common product output is concerned, this must be taken as the additional output over that which can be supplied by the original project. Whilst the performance of the original project can give a guide to the likely performance and some of the costs of the new project, the output required of the additional capacity, to meet demands, will require careful assessment through market research. The results of this research can influence the capacity needed of the new project and also the performance needed from the new asset.

If the original and new projects are separated by a number of years, the new project may well utilize updated technology which could influence both operational and maintenance costs, hopefully for the better. In such a case, the economic assessments carried out for the viability of the project can differ considerably from the assessment carried out to determine the optimum operational regime that should be adopted, shared between the original and the new assets.

As stated earlier, the viability of the new project must be based upon the output needed to supplement the original project to meet the demand for the product. However, should the new project be more cost effective to operate and maintain per unit of product than the original, the optimum way of meeting the demand will inevitably require greater output from the new project at the expense of output from the original. The sharing of loadings will further influence the balance between the acquisition costs and the required performance from the new project and this should be taken into account for the new specification.

Other aspects which may have changed since the original project, relate to the financing of the new project. This could include any grants that are offered for the acquisition of the assets thereby affecting the net acquisition costs. The materials used either as feed materials for the product or to service the asset or the production process may well have changed. This could, of course, affect both projects and may well have influenced the marketable prices for which the product can be sold. Finally, the source of the finance for the new project may be different from that obtained for the original. Even if the same source is to be used, the return required on the borrowings may have changed as circumstances such 'bank rates' alter and this will normally affect the 'discount rate' to be used in the economic evaluations.

Risk and sensitivity analyses

For all new projects there is invariably a measure of uncertainty about all the data regarding both costs and performance of the asset. This being so there is bound to be a degree of risk that the new project will not, in practice, prove to be economically viable. When carrying out economic assessments of a proposed project it is useful and recommended that a sensitivity analysis be carried out which, for each input factor to the assessment, identifies the limit to which that factor could vary before the project fails to remain viable. The results of this analysis need to be compared with the results of a separate consideration of each factor to identify the likely range of values it might take in practice. This comparison will help to identify the risks involved and areas in which particular care must be taken.

Projects for public bodies

As far as public bodies, such as the government or charities, are concerned, most projects are initiated by a perceived need by the public or the beneficiaries of the charity. Such projects are seldom based upon achieving a return on the moneys deployed in the project. Nevertheless, economic analyses should be required for such projects. The objectives of the proposal should be clearly established and should form the basis of the (ideal) performance specification. Meeting this specification should normally be a formal requirement of all tenders for the provision of the required services. Thereafter the economic assessments will be essentially to establish the minimum life cycle costs for the project. Not all the specified requirements will necessarily have the same priority, so there may be scope for considering variations in extent from the formal specification. To do this, some arbitrary measures of valuation against performance variations must be used. These are best established by the user in advance of receiving offers or tenders and then used in comparing the offers of potential suppliers. Circumstances will dictate whether these valuations should be revealed to the potential suppliers. To a large extent, the acceptability of variations from the 'ideal' specification will depend upon circumstances, not least of which is the availability of finance for the project.

Chapter 13

Performance Specifications for the Asset

The preparation of the specification for the asset and its components is the responsibility of the Project Manager at the start of the 'acquisition phase'. An outline of the proposed asset and its use in producing the product will have been established in the 'concept phase' by the sponsor. This will have been carried out in some detail, sufficient for reasonable estimates to have been made of costs and output for use in the economic assessments submitted for the project's approval.

Project Manager's responsibilities

As described in earlier chapters, the Project Manager should seek inputs from all affected sections of the user organization, for inclusion in, or for consideration for inclusion in the specification. Unfortunately, in seeking ideal conditions for their own individual areas of future involvement, some of these inputs could be mutually incompatible. To resolve these incompatibilities, the Project Manager will have to establish compromise positions. These positions should clearly be established as those which give the best economic results for the project as a whole over its life-time.

Typical conflicts in requirements

An example of one such conflict of ideals often arises between the operators and maintainers of the proposed plant. Half a decade ago plant was usually controlled by manually operated devices fitted directly on the plant itself. Nowadays it is to be expected that controls of these devices are brought to a common central control point and operated by remotely controlled electrical, hydraulic or pneumatic actuators. This would, today, be a minimum requirement often dictated by the plant size and the physical efforts needed to operate the devices. More importantly it reduces the number of operating personnel required, thereby significantly saving operating personnel costs. In many cases these days, devices will be controlled automatically and without operator intervention. This reduces the operating personnel requirement even further. On the other hand, the maintenance staff will seek for the plant to be as simple as possible to minimize the future maintenance costs. The introduction of remote controls, actuators and automatic controllers adds to the maintenance work load and consequently the maintenance manpower and costs. A compromise situation must therefore be established. In doing this, the Project Manager must take into account the rapid changes in technology in this area of controls, instruments and automatics which may alter or displace the optimum point of the compromise in the future.

The above example is only one of the balancing acts that the Project Manager will have to face in drawing up the specification and in particular the performance requirements to be contained therein.

Project's capacity requirement

One further compromise concerns the asset's availability and its capacity in terms of its rated output. Inevitably the project will be based upon a target output of the product which was declared at the time of the project approval. This target may not be a constant rate but will be dictated by the likely demand from 'customers', an input to the proposal based upon the estimations of that section of the organization tasked with carrying out the required 'market research'. To meet this demand, the capacity of the asset must be set at a level that, allowing for its availability, will be able to satisfy the peak value of demand.

Effect of operational management techniques

Modern thinking on operational management tends towards the 'just in time' philosophy, thereby avoiding the need for storage for raw materials, between stages of production or of the finished products. This also reduces the inventory and cost of work in progress. However, if the demand is likely to be highly variable, the peak rated capacity of the asset may be capable of being reduced and its acquisition costs reduced. The peak demand by customers can then be satisfied by a degree of product storage or, in some cases, by inter-production stage storage. Where demand is highly variable some degree of storage may be inevitable, but the size of the storage and its overall costs should be optimized against the savings in costs associated with the reduction in specified capacity of the asset. The other factor to be taken into account in this optimization exercise is, of course, the availability of the asset. It is obviously essential for both the capacity and the required availability of the asset to be established in the specification. Recognizing that these are critical to the viability of the project, these two parameters should be established as minimum values to be guaranteed by the supplier. If this is not deemed possible in certain circumstances, target values should be set at the very least. Care must be exercised not to require an availability performance level which is too high. The assets may be capable of being designed to achieve higher availabilities, but as the target levels are raised, so the acquisition costs are also likely to increase and the law of diminishing returns will inevitably apply.

Options

A series of options based upon varying values of capacity, availability and storage need to be devised which are capable of meeting the output demand authorized for the project and which meet the likely customer demands. These should then be assessed for their life-time costs and incomes to determine which offers the best economic optimum solution.

Balancing of costs

There are three major areas where the specification must be set such as to establish the balancing point between conflicting requirements. In effect the above discussion includes a prime example of one of these. These are (a) the balance between 'performance' (availability in this case) and initial acquisition cost, (b) the acquisition cost and its balance point against running costs, and (c) the choice of technology. Selection of components based upon a fully developed technology are likely to create lower costs in acquisition and in some aspects of maintenance than the use of the latest, and perhaps not fully tried, technology. In addition to cost estimates the risks associated with them are likely to be lower or at least better understood than those of the latest technology.

Energy requirements

Most processes require an energy input. Even a welfare office will have demands for lighting, heating, air conditioning and office machines. Here the latest technology is likely to achieve the requirements with greater energy efficiency. The cost implications of this may or may not be significant. However, the possible rise in energy costs as indigenous resources are exhausted, as well as the environmental effects of waste heat and the by-products of extra fuel usage must be considered. In this respect some new technologies may show improvements over more mature alternatives.

Materials

Another aspect which might need guidance from the users to the suppliers can relate to the materials to be used in the assets and components. This again is dependent upon technological developments and will vary from item to item. Choices for some items may have to be made between commonly available materials of low specific costs and more sophisticated materials at a higher cost. The latter may offer longer useful working lives and the prospect of reducing the need for replacement during the component life. In these cases the reduced maintenance costs can justify the extra initial cost of using the more sophisticated materials.

Maintainability

The maintenance department will have contributions to make to the specification, and certainly to the subsequent evaluation of the tenders, in terms of accessibility and maintainability of the plant. These aspects, frequently ignored by designers in the past, are now becoming more understood and with the introduction of international standards on these topics can be readily dealt with in the specification.

Duplication of components

From the operator's point of view, the availability of the plant overall may be principally controlled by the availability of specific components. The merit or definite need for duplication of specific components should be considered in terms of the overall economics of the options.

Economics – basis of choice

The specification must, as a first priority, establish the user organization's requirements to satisfy the established objectives of the project. It should also contain any other requirements that the user deems essential and for which alternatives would never be acceptable. The discussions early in this chapter deal with the most important aspects of the first two considerations. It has been said many times that the choice of options in all areas and especially in the design of the asset should be based upon the economic optimum in relation to the project as a whole over its life-time.

Carrying out the economic assessment of alternatives to establish the optimum for the user organization may well involve data, costs and values known to the user but not the supplier. For commercial and other reasons it may be inappropriate for the user to reveal these externally. In such circumstances, the user should undoubtedly give guidance to suppliers where his assessments show a marked preference for a particular course of action or an optimum point of balance between opposing trends. It is seldom that a user can act more directly as in the case of the electricity supply industry where tenderers for transformer contracts are advised of the assessment penalties to be imposed according to the offered transformer's energy losses. Nevertheless, internally the user organization should establish, for the use of the project manager, a suitable schedule of 'penalties' to be used in the assessment and comparison of tenders which fail to meet the specification's requirements or guidance.

Quality control

Quality and quality control procedures will inevitably require some inclusions in the specifications. The universal establishment of ISO 9000 series of standards, (BS 5750 in the United Kingdom) can form a suitable basis for these aspects. However, these may need to be supplemented by certain specific requirements or modification in the light of the nature of the asset being acquired.

Strategic spares

Certain assets, particularly those containing plant and machinery may warrant being supported by strategic spares for particular components. Where these can be identified at the specification stage, it is usually beneficial to include them in the specification as optional extras. By so doing the user tends to prevent the supplier from quoting a low price for the main asset or component and then to recoup his profits by asking a high price for spares and replacements at a later date. A further merit in ordering such spares early is in gaining their help in ensuring that the commissioning dates will not be delayed by failures during commissioning. There can be further advantages and these are described in further detail in my other book, published by Mechanical Engineering Publications, *Strategic Spares and the Economics of Operations*.

Programme and penalties

For most projects the commissioning of the asset by a specific planned date is essential if forecasts of costs and income, made in economic assessments, are going to be met. Late commissioning can have a serious impact on the viability of the project as a whole. In the circumstances the user should identify his planned commissioning date in the specification and require the supplier to confirm that this date will be met. At the same time it might be appropriate to identify the penalties to be imposed if the date is not met. Furthermore, the specification should require the supplier to give details of his proposed programme against which the contract can be monitored.

Maintenance periods

Finally, most acquisitions will need to be covered by the supplier against failures for an initial period, known as the maintenance period which should be identified in the specification. This normally covers the costs of materials and labour used in rectifying any component or system faults. Beyond this period, certain guarantees on specific items may be appropriate, usually at extra cost. The supplier should be asked to state what he is prepared to offer in that respect; these offers will need to be assessed in the tender comparisons and the post-tender evaluation of the project.

Chapter 14

Design of the Product

The specifications

In the last chapter we dealt with the specification of the user's asset. It was written in terms of the user preparing the specification to ensure that his requirements were properly identified to, and tenders requested from, potential suppliers. By implication this was written on the basis of a prospective contract or series of contracts between a major user organization and supplier organizations for a new major asset (a ship, oil rig, or food processing plant etc.) with which to produce his own product (transportation of cargoes, crude oil or packaged foods).

A 'smaller' user will still need to draw up a specification for the asset for his own new project. However, this may be used internally, rather than issued to suppliers, as a guide to the selection of components from those already being produced and marketed.

'Open' market products

As a corollary to the above, for a project to produce an output in significant quantities to be offered for sale on the open market, the supplier organization will act as user organization in respect to his production asset. At the same time he will be responsible for the design of his product whether this is a special 'one-off' to be offered to a user organization in response to the latter's specification and request to tender, or a mass produced item for the 'open market'. In this context an open market may be one serving the general public or a more specialist one such as those purchasing large air compressors. Whilst there are clearly major differences between designing products for an open market and those of the one-off variety, there are common factors and equivalent methods of dealing with these.

The first of the common factors, common to most industrial projects, is that the supplier's product often becomes a user's asset and the user's product becomes his customers' assets, and thus a chain of products/assets becomes established. The principles of economic management should therefore be applicable to both the supplier's production assets and to his products. Whilst the first is based upon the economics from the supplier's point of view as a user, the products need to be viewed from the point of view of their users.

As well as viewing his product from the customers' economic point of view, the designer should take into account the whole of the customers' requirements. In the case of the one-off product, the user/customer will have established his requirements, as originally perceived, in his specification. In the other case where the product is intended for a multi-customer mass market, each potential customer may well have slightly differing needs. In general these will not be submitted to the supplier in the form of a formal specification. Thus the supplier must take action himself to establish his customers' needs. His design of the product can then be aimed at satisfying these needs to the maximum degree possible.

There are a variety of ways in which the designer can obtain information regarding the potential customers' needs. These include questionnaires sent to actual customers and taking note of prospective customers' queries when seeking detailed information about the product before purchasing. As the latter may have to be transmitted via retailers and agents, such information may be more difficult to collect and could become distorted in transmission. Ordinary market research methods may help to establish certain common requirements but the questions posed to potential customers would need to be carefully controlled and the extent of aspects to be covered is likely to be limited. A somewhat wider selection of questions, covering more of the areas needing to be researched, could be given to a more restricted panel of typical prospective customers invited to sample the product for a suitable period of time and then answer a questionnaire. Reimbursing or rewarding the panel could motivate them into giving useful answers.

An outline of the product design process is given in Fig. 14.1.

Standards

If the product is of a type that is covered by a standard, usually a national standard, such a document may well contain the views of users. Even if the

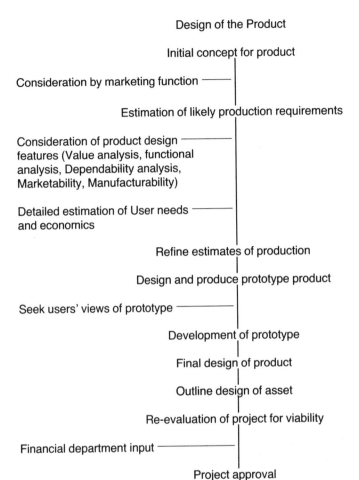

Fig. 14.1   Development of product design

views themselves are not recorded, the product standard will probably contain recommended design features which have been formulated as a result of users' comments. In addition to the foregoing, such standards will also contain features which relate to the compatibility of products produced to the standard. These features may also be of benefit to the user, although not specifically expressed. This can arise in the event of the user requiring to replace an original component, only to find that the original source no longer exists. This can easily happen as a result of take-overs, closures or just obsolescence. Another feature of standards is that their latest versions will normally cover aspects of product safety and the national interpretation of national regulations or EU Directives.

Value analysis

Armed with the results of these research methods, described above, the proposed design of product should be subjected to both value and functional analysis. More detailed information about these techniques is due to be published shortly in a European standard to be issued by CEN (Comite Europeen de Normalisation). A classic example of the use of value analysis, often cited, relates to a manufacturer who designed his product to be sold in a special protective case. It was then discovered that the majority of his customers removed the product from the case, which was then discarded, and fitted it as part of their own product.

Quality requirements

It is axiomatic that the user will be concerned with the quality of his asset as a whole and also its various components. The aspects of quality will include such matters as appearance as well as the various performance characteristics. The degree to which the user will be concerned and willing to be involved will vary from component to component. This will be determined in part by the nature and complexity of a component and how critical it is in the operation of the asset. Clauses in the specification should cover the quality requirements of the components and of the whole, reflecting the degree of concern felt by the user. This may vary from simple checking and agreed control limits, to a need for extensive quality management and procedures. These latter, should if appropriate, be based upon the ISO 9000 series of the International Standards Organization's standards. These standards are replicated by both the United Kingdom's BS 5750 and the European standard EN 29000. The degree of adaption of these should be tailored to the item and the requirements by selection of appropriate clauses from the standards. When the requirements are extensive, it may be appropriate for the standard to require, or at least encourage that suppliers be certified as being in conformity with these standards.

Dependability

Another similar aspect that may be covered in the specification relates to the user's requirements regarding the 'dependability' of the products to be supplied in his asset. The term 'dependability' is now used in international standards to cover the aspects of 'reliability', 'maintainability' and 'maintenance support'. The close relationship between 'quality' and 'dependability' management is reflected in the fact that the management of dependability is now covered by a further part of the ISO 9000 series of standards (ISO 9000–4). This is a dual standard issued both by ISO and the IEC (International Electrotechnical Commission).

Maintenance support

Regarding maintenance support of products, the designer should ensure that any need for such support is properly conveyed to the user. For major plant and contracts, the user should indicate to the supplier what facilities for supporting his asset he already holds. This is normally done via the specification and includes such facilities as central workshops.

Manuals

In Europe, it is now a requirement under the 'Machinery Directive', EU 89/392 EEC that products should be supplied together with a manual giving certain essential information to the user. For major contracts it is likely that the requirements of manuals is extended to every component of the asset. Guidance on the preparation of these technical manuals is given in a British Standard BS 4884. This was recently revised and now Part 1 covers aspects which are deemed essential to all manuals, whilst Parts 2 and 3 are guides to 'contents' and 'presentation' aspects.

Chapter 15

The Acquisition of the Asset

Acquisition policy

The first stage of the acquisition phase is the drawing up of the specification(s) for the asset as discussed in Chapter 13. During this period, two further activities should be taking place. Firstly, there should be a consideration of and a decision made on the acquisition policy to be adopted. In particular it should be determined whether the full asset should be dealt with in a single 'turn-key' contract or alternatively should certain parts be dealt with under separate contracts between the user and specialist suppliers. For smaller projects, involving no new technology or components, it may be more suitable for individual components, already available on the market, to be purchased separately and erected by the user or under a separate contract. Variations, combining different elements of these procedures are possible and the final choice should be determined by the nature of the asset, its components and the circumstances. In this determination, the economics of the various alternatives should be considered but other factors such as the ability to satisfactorily and reliably coordinate different suppliers' activities may prove to be more influential. Evaluation of such factors as these is very difficult and thus a full economic assessment of alternative policies could lead to doubtful conclusions as to how to proceed.

Figure 15.1 gives an outline of the process for the asset design within the user organization.

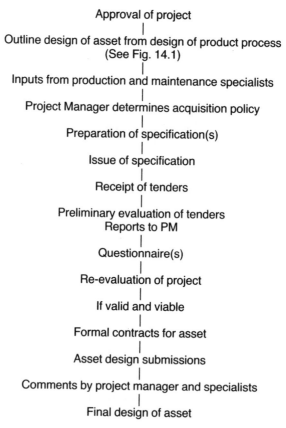

Approval of project
|
Outline design of asset from design of product process
(See Fig. 14.1)
|
Inputs from production and maintenance specialists
|
Project Manager determines acquisition policy
|
Preparation of specification(s)
|
Issue of specification
|
Receipt of tenders
|
Preliminary evaluation of tenders
Reports to PM
|
Questionnaire(s)
|
Re-evaluation of project
|
If valid and viable
|
Formal contracts for asset
|
Asset design submissions
|
Comments by project manager and specialists
|
Final design of asset

Fig. 15.1   Outline of development of asset design in user organization

Vendor assessment

The final decision on acquisition policy will determine the way in which the specification or specifications are drafted for submission to potential suppliers and/or the purchasing/contracts sections responsible to the Project Manager for this project. It will also dictate the way in which the second parallel activity is handled. This second activity is directed at the drawing up of a tender list for each of the contracts that have to be let and an approval list of potential suppliers for items to be directly purchased. The incorporation of a potential supplier should only take place after the organization has been found satisfactory in a vendor assessment exercise carried out by or for the user organization. The nature and degree of this approval exercise will depend upon the circumstances and the extent and criticality (in the project) of the contract/purchase. The factors included in this vendor assessment exercise could include any of the following. Firstly, the 'stability' of the supplier organization should be found to be adequate. The 'stability' should include its financial stability such that there is an assurance that the organization will remain functioning for the entire duration of the contract, and not become bankrupt. The possibility of firms being taken over is generally high at the present time. The existence of a take-over bid or the suspicion that one might be in the offing may not automatically rule out a particular firm. This will mainly be determined by the reasons for the take-over. If these reasons do not threaten the acquisition contract and continuation of production and supply appear assured, their inclusion in the tender list is probably acceptable. However, where major machinery items are concerned, the user may have a need for the manufacture/supplier to undertake some forms of maintenance at intervals throughout the project's useful life. In such circumstances as these, a much longer view on the 'stability' of the supplier may have to be taken.

A second aspect to be considered in the vendor assessment relates to the financial resources available to the supplier for meeting the needs of the proposed contract. This is in addition to the general question of the financial stability discussed above. A third important aspect is the experience of the supplier to carry out the work proposed. Has he done similar work in the past and for whom? Confirmation of satisfaction might need to be obtained from previous customers. Has he the necessary tools and machinery and the technical backup to carry out the work? Finally, the general reputation of the firm in respect of earlier contracts that it has undertaken could influence the decision whether or not to include it in the tender list.

Purchases

After the above preliminaries have been completed, the normal contract process can go ahead. For the simpler items to be acquired by purchase procedures, those suppliers approved for these acquisitions can be approached for information about their products, needed in the project, as identified in the purchasing specification, and prices.

Contracts and tender assessments

For major items to be obtained through contracts, suppliers on the tender list should be sent a copy of the relevant specification and requested to submit a tender by a specified date. When the tenders are received they should be subjected to a preliminary assessment. This should be carried out by the Project Manager and his team, but also by representatives of the operations and maintenance departments and all other departments who are affected by or involved in the project. Tenders for major contracts seldom completely conform to the specification. The review of the tenders should identify deviations from the specification and any aspects which, although conforming, have features which are deemed not completely desirable. These points should form the basis of appropriate questions for a questionnaire to be sent to the tenderer. At the same time the specialists should estimate the additional costs of rectifying the deviations or their implications on costs and income during the useful life phase if they were accepted. These should be sent to the Project Manager for inclusion in economic assessments of the whole project if that tender were adopted.

Questionnaires

This preliminary assessment may well rule out certain tenders either on consequential project economics or content or both. Clearly questionnaires should not be sent to these tenderers but to all other tenderers who appear 'in with a chance'. The suppliers' replies to their questionnaires should give priced options for modifying their original tender. The acceptability of each option should be carefully considered both from the technical and life-time economics points of view before being deemed to be included in the modified offer. Selection of the preferred tenderer should then be made on the basis of a full life-time economic appraisal of the project as it would be if each modified offer were incorporated. These appraisals will also incorporate the relevant specialists' evaluations of deviations and design weaknesses, revised as necessary by the replies to the questionnaires.

Reassessments of project economics

The results of the economic appraisal of the preferred tenderer's modified offer must also be compared with the economic appraisal carried out for the project's authorization and for general viability. Should this exercise show that the viability no longer exists or has been seriously diminished, a re-submission of the proposal for the project should be made to its original authority before proceeding any further.

Contract letting

If the project is to go ahead, the next stage is for the Project Manager to prepare and negotiate with the supplier the terms of a contract for the asset or component concerned. In preparing the contract a number of points need to be considered for incorporation if these have not been explicitly covered in the specification/tender and questionnaire answers. Some of these will be discussed below. They are based on the author's own experience which has been principally concerned with large, complex plants whose expected lives are measured in decades rather than years. For lesser or shorter lived projects these may not all be relevant, but it is advisable for these to be at least considered.

Contract extent

For long lived projects the following aspects are particularly important. Firstly, the contract should unambiguously define its extent. This should be in terms of which of the following functions are included, the supply, its erection or installation, its testing and commissioning and its support. It should also be in terms of ancillary items of supply. These include Technical Manuals, test equipment, special tools and rigs and any staff training needed. Even if the contract does encompass the erection/installation by the supplier, full instructions for these operations should be included, preferably in the Technical Manual, in case dismantling and re-erection is needed in the future. In the past the Technical Manuals have generally been of poor quality and seldom met all the user's requirements for information or instructions. Only where the manuals were required to be submitted to a statutory body, such as the Civil Aircraft Authority or the Nuclear Installations Inspectorate, did they approach the standard now deemed essential. Essential requirements and guidance on contents and presentation are now contained in the latest (1992/3) versions of British Standard BS 4884. Training of both operating and maintenance personnel is usually a user's responsibility, but when novel technology and designs are used, some training of the initial complement of staff by the supplier/manufacturer's personnel may be required. The need to train replacement staff in due course must be remembered both from the point of view of allowing for the cost of such training to be included in the economic appraisals but also as to whether this will still need the supplier's help or whether, having gained running experience, this could now be carried out 'in-house'.

The management of quality and also dependability during the contract should be laid down in the contract itself. Reference to the standard series

ISO 9000 or its equivalents could be an effective way of doing this. This section of the contract should also cover the procedures to be adopted for testing and the witnessing of tests by the user's staff or representatives. Certificates of certain tests may also be requested and for these to be submitted to the user's Project Manager.

Spares

A second major area concerns the provisioning of spare parts. The advantages of obtaining strategic spares for major components, under the contract, has been discussed earlier. However, there will also be a need for routine spares which will be needed throughout the useful life of the asset or component. These are likely to be stocked by the user in limited quantities determined by an economic assessment of their acquisition process. To establish the need for such spares and to determine the timing and quantities for initial and subsequent re-orders, an input from the supplier is generally needed. A request for a priced Recommended Spares List, which also gives lead times and anticipated usage rates should be included in the contract. For long life projects the continuing availability of spares could be crucial. On the other hand, asking the supplier to guarantee to be able to supply a particular spare for the full length of the project may not be reasonable. However, the period for which the supplier if prepared to make such a guarantee should be covered in the contract. At the same time the user will need as much notice as possible if any of these spares are to become unattainable from that source. To cover this eventuality it is recommended that the user places clauses in the contract requiring the supplier to give such minimum declared notice and, if the user so wishes, to supply drawings, material specifications and sources and manufacturing information for any item to be discontinued. This would then allow the user to manufacture or to get a further supplier to manufacture further replacements for his continuing use. This aspect is discussed more fully in my earlier book – *Strategic Spares and the Economics of Operations*.

Completion date and programme

As completion of the asset can be critical to the project as a whole, the user will need to monitor progress of the design and production processes for each component and delivery and erection of the whole asset. This should be monitored against a programme to be drawn up by the supplier and to be approved by the Project Manager. For large contracts this may need the user's staff or representatives to be able to physically check progress at the manufacturers' works. Agreement to this procedure and the granting of access for the purpose should be covered in the contract. If relevant to the contract any progress of design submissions and approval should be covered as well as the physical progress of components through their manufacture and supply.

Guarantees

Finally, it should be established that any guarantees against failure etc. of a component should start with the agreed completion date for commissioning the asset. It is unacceptable to users to find that a minor sub-contractor has set a date for the guarantee on his parts to start on the delivery date to the main contractor only for this guarantee to expire before the user has been able to use the asset for his own production. This can be a problem where long erection periods are involved.

Chapter 16

Operational Aspects

Need for early involvement

By the time that an asset is commissioned and production can start, the design of the asset will have been fixed in a way that will determine the majority of the useful life operation costs. For this reason, the operational staff must get involved at a much earlier stage in order to influence their operational costs in service. One example, relating to the extent to which automation has been incorporated in the asset's control systems, was discussed in Chapter 13. The layout of plant and its controls can also have a significant influence on the operational manpower required and consequently on the operational labour costs. Full attention of operational specialists at the specification, tender assessment and design submission stages is merited and should ensure that these 'committed' operational costs are kept at or near the optimum for the project as a whole.

Operational plan

As the asset is being brought to readiness for operations, the sales or marketing departments should determine a plan for supplying the product to its market. In the case of public services, an equivalent department will determine the output of product needed. The operational staff must then translate this into an operational plan that will enable this supply plan to be met. In deriving this operational plan, allowance must be made for the needs of maintenance of the asset. This will often entail having to stop or restrict production from the asset at intervals. These intervals must be co-ordinated with the maintenance department and timed to coincide with periods when the operational plan has achieved a sufficient output of product over that called for by the supply plan. The loss of production during the maintenance downtimes can then be made up by the prior excess production and storage of product. For products in which a continuous supply of product is called for, such as electricity, gas or water, either the project itself must be one of a number providing an output of product or, within the asset itself, there must be a number of parallel production streams. In either case the asset or assets should have a total capacity with sufficient redundancy to allow for the maintenance outages.

Operability

The 'operability' of the asset will have been mainly set during the design and erection stages, ideally after inputs from the operation specialists. However, after being in use for some time, some operability difficulties may be found. The solution of these difficulties may be to carry out minor modifications to the asset. These modifications should be carefully planned and carried out at a time when the asset is shut down for other purposes such as essential maintenance.

Fault identification

One of the most difficult of the tasks facing the operators of plant and equipment is the identification and location of faults when they occur. When a fault occurs with the plant in service, the operator's first task is to return the plant to a safe state. Concentration on this activity, which might take a significant time, often leaves very little time for attention to be devoted to recording the symptoms of the fault. Then, once the plant has been rendered safe, usually shut down, the symptoms disappear. With modern complex plants this can be overcome, at least in part, by computers recording plant system readings and the initiation sequence of alarms being raised. These may not fully identify the source or nature of the prime fault, but analyses of the system as part of the design studies can help in locating the most likely sources. The carrying out of the processes of Fault Tree analysis (FTA) and Fault Mode Effects and Criticality Analysis (FMECA), during the design stage is principally aimed at identifying potential problems which may need design attention. However, the results of these analyses, possibly redrafted into a more usable form, can assist in speeding up fault identification and location. Fault location tables and procedures should be incorporated in the Technical Manual for the asset. Guides for the carrying out of the two processes of FTA and FMECA are now published by the International Electrotechnical Commission in their publications IEC 1025 and IEC 812 and now, in slightly extended form in British Standards' BS 5760, Parts 7 and 5 respectively.

Emergency actions

As was mentioned in the last paragraph, in the event of a fault the operator's primary task is to bring the plant or equipment to a safe condition, often the shut down state. This may involve unusual emergency actions. Such emergency actions should be given in an emergency procedure in the operator's instructions section of the technical manual. However, the exercise of these procedures should be a rare event precipitated by the infrequent faults. In normal operation, opportunities to carry out these emergency procedures could be rare and thus not completely remembered when they are called for by a fault occurring. As a consequence a measure of training of the operational personnel in these emergency procedures is essential. Live practice on the plant may not be possible for various reasons; loss of production, or potentially causing damage to components of the asset are common reasons. In such situations, alternative training methods have to be devised. These may involve 'dry runs' on the shut down plant or, in extreme cases, such as aircraft or nuclear reactors, simulators may be necessary. These have the particular merit that extremely serious fault conditions can be simulated and 'experienced' by operators. This would include multi-engine failure in an aircraft or highly positive temperature coefficients in a thermal nuclear reactor.

Hazards

Apart from the plant failures discussed in the last paragraph, the operating personnel have to be trained in dealing with other more common hazards such as fire or major leakages of fluids from the process or auxiliary plant. These may call for written procedures and practice drills by the staff at reasonable intervals. The regulating authorities may also be involved.

Environmental aspects

During the useful life phase, the impact of operating the asset on the environment must always be watched by the operating staff. In certain cases, usually the subject of legislation or statutory regulation, corrective action may have to be taken urgently by the operators. This may extend to a point where the asset has to be shut down. Operations may impact on the environment in many different ways. At one extreme the asset itself, or the production process it is designed for, may impact directly on the environment in normal operation. Excessive noise from plant is one such case of this; more likely are the effects of operating the asset either at extreme limits of its designed capability or under abnormal conditions. When heavy plant is started up, especially boiler plant, discharge of drains from steam lines can result in high noise levels as well as the discharge of water and steam. Similarly, in other process plants there may be similar discharges from vents from which even more noxious fumes may be released to the atmosphere. At the other end of the production range, when plant is attempting maximum output but under abnormal conditions or worn plant, 'normal' discharges may become excessive. Using the boiler plant example again, the flue discharges may increase to a point which exceed statutory targets, if the boiler is fired with fuel of quality below that specified at the design stage. If mechanical plant becomes worn, the quality of the output may become so affected that the process produces a large number of rejects or waste which can cause environmental problems in its disposal.

When the basic design of the asset or its operation is the cause of environmental impact, redesign is the only long term remedy. Normally this involves the incorporation of modifications to the plant or equipment. When the problem arises from feed materials, fuel etc. not being to the standard or specification assumed when the plant was designed, modification may allow the process to be brought within the acceptable limits of impact on the environment. Alternatively, a review of the design and of the project as a whole, may suggest that a degree of de-rating of the asset may be required in addition to any modifications. In all these cases where modifications are deemed necessary, the cost of the modifications and the impact on the economics of the project over the residual life should be assessed and compared with other possible alternatives up to and including the shutting down of the project forthwith.

When the problem arises from wear, the correction must emanate from maintenance of the affected plant or equipment initially. This should be accompanied by a review of the maintenance requirements of the plant etc. and revision of the maintenance plan and, as a consequence, of the operational plan also.

Operators' responsibilities

Although the corrective actions to combat environmental problems are seldom the operational staff's direct responsibility, operators remain the staff responsible for recognizing that a problem exists and its probable cause. They are also responsible for taking any immediate actions to keep the asset operating within statutory or agreed environmental limits even if this results in deviation from the production plan. The cost of penalties for not doing this can often exceed the cost of loss of production until a rectification is carried out.

Management of feed materials

The management of the operational processes should be optimized as far as possible. This will include the supply and flow of feed materials and essential commodities used by the asset, including energy. Techniques such as JIT can and should be exploited as far as economically justified. Arrangements with suppliers to provide materials etc. in a way that matches the demand on the asset, can be very cost effective. However, if a supplier is so closely tied to the user organization's requirements that he has virtually no other customers for a particular product, care must be taken to establish his capabilities. Attempting to contractually place the responsibility for maintaining the required flow of feed material 'just in time' may be dangerous. Any supplier organization can be affected by plant breakdowns, transportation problems or industrial action. The impact of these and the consequential interference with the flow of feed materials to the user's project may be more serious economically, for the user organization then the supplier. As a result, if there is to be such close coupling of the supplier's activities with those of the user, a joint assessment of the supply/usage train and the risks involved could be advantageous. The effects of a buffer stock of this feed material should also be assessed in terms of moderating the risk described. These common assessments should be made on the basis of the effects on the user organization primarily but also on the supplier organization. These assessments should form an element in the establishment of the contract between the two parties.

Other management techniques may similarly offer the possibility of improving the costs of the production processes. As in the case of the JIT technique described above, these should be adopted as far as possible without going into stages in which the cost of developing them further would exceed the financial benefits that they achieve. In this the evaluated cost of risks may be taken into account in a similar way to that described in the above paragraph.

Procedures

In devising the operational procedures the application of ordinary operational research may be appropriate and beneficial. These should not only review the activities of the human operators but also should be applied to the individual processes, with the aim of minimizing waste, rejects and the need for reworking.

Quality control

The degree of quality control and the extent of quality management will greatly depend upon the circumstances and the processes involved. The arrangements described in the international standards of the ISO 9000 series form a reasonable guide for any large complex process. However, many small organizations claim that the standards are too involved and expensive for their more limited circumstances. Nevertheless, the standards can reasonably be taken as a guide from which elements can be adopted if they are appropriate and relevant. An important element of the quality chain should always be the proper receipt and treatment of the feedback from customers.

Monitoring the asset

The performance of the asset and its constituent elements will need to be monitored for most of its useful life. Each element should be monitored for its throughput and the quality of its output. The energy input and usage of service materials, such as lubricants, should also be recorded. Each of these aspects should be compared with the 'standard' values either obtained during commissioning tests or from design predictions. Variation of these parameters from the 'standard' figures could give warning of wear taking place or that damage has occurred. Apart from generally causing extra costs in operation, they may give early indications of the need for maintenance which can then be planned to be carried out at the earliest opportunity with minium cost implications. When energy consumption forms a significant element in the running costs, the sources of energy loss should be sought and action taken to minimize these as far as practicable in the circumstances. The performance of energy recovery devices should also be checked. Whilst the performance in this respect of plant will form the major area of consideration, the thermal performance of buildings will also warrant attention. These tests of performance may be achieved by continuous measurements or by spot checks at intervals of time, as is most appropriate. If the latter, a proper programme of tests and procedures to be followed, should be devised for the operational staff.

The collection of performance data has always been the responsibility of the operational staff, but was often disregarded if other urgent operational tasks arose at the time that readings should have been taken. The use of automatic recordings of instrument readings has, to a large extent, overcome this problem and is more likely to reveal the origins of problems in the asset. Retaining records of readings for future analysis can also reveal the effects of long term trends and deterioration which could assist in planning or revising the planning of maintenance tasks. Reports on incidents involving the asset are a further source of information, not only for identifying maintenance requirements but also for consideration when further assets are being planned or modifications made to the existing asset.

Safety

In general the operating staff are in the front line when it comes to the safety of personnel and equipment. In this respect the personnel includes anyone who has any contact with or who has to approach the asset. Where plant or equipment are concerned, the operator is normally responsible for isolating the plant from all dangers before it is handed over for maintenance or testing. The operating staff will have to discharge any hazard such as pressurized fluids or electric charges that could threaten the wellbeing of personnel. In some cases mere access to certain areas may introduce risks such that access control, as well as the safety control over the proposed activity, is necessary. High radiation zones in a nuclear plant is an example of this. In considering the operational manning of the asset the time and effort required by these safety activities must be taken into account. Additionally the facilities needed to carry out these safety duties must be provided. This can include special test facilities such as a high voltage probe or even extra staff to carry out surveys and measurements such as a radiation monitor or a person checking on personnel working in confined spaces.

Emergency planning

Some assets may contain potential hazards which pose a threat, not only to the asset itself but to anyone or anything in the vicinity. This may well include the public at large within a certain radius of the asset. These hazards together with the nature of the risk and its extent should be fully assessed as a prime task of the operators. From these assessments emergency plans should be drawn up and discussed and agreed with all the outside authorities who need to be involved. This may well include the fire and rescue services, police and local government departments. A requirement for such emergency plans has been statutorily placed on nuclear sites and some other plants. Although not always required by law, similar plans for other hazards may be called for by local government: as environmental risks assume greater importance, the desirability of such emergency plans will increase and become subject to searching criticism. An early start on such tasks, even though not yet demanded, should enable most of the potential criticism to be identified and considered and, if necessary, adjustment can be made to the asset or the way it is operated. Training for these emergency plans should be given and, since they will seldom be activated in earnest, retraining will be necessary at intervals. In the case of the statutory plans of the nuclear sites these have to be exercised at intervals and a degree of exercising of voluntary plans, if only to test their feasibility, is normally warranted.

Routine servicing by operators

Although strictly a maintenance activity, most lubrication tasks are often delegated to the operational personnel to carry out on a routine basis. The routines to be followed in this respect should be devised jointly by the operators and the maintenance staff and should normally be based on the wear discovered in the periodic examinations of the components involved by the maintenance staff. Other minor maintenance tasks, such as adjustments that can or must be carried out with the asset in production, may also be delegated to operational staff. However, in these cases arrangements should be made to record such operations in the maintenance/plant histories so that they can be taken into account in any future plant reviews.

Economic basis for choice of procedures

When alternative operational plans or procedures are being considered, as in all such cases, their effects upon the remaining life-time economics of the asset should be compared. For these realistic values of the key financial inputs, as opposed to the initial assumed figures during pre-operational phases, should be used. Firstly, the discount rate must reflect the cost of financing the variation in the project plan involved as well as the rate of return needed on such expenditure. This may be different to the discount rate used in evaluating the project as a whole in the concept and acquisition phases. Another factor that might have significantly changed since the

Table 16.1 Summary of contributions from operational/production specialists

Concept Phase
(1) Consider the sponsors outline proposals for the asset for the new project and assist him in estimating the staffing requirements, labour and running costs and its performance.
(2) Examine and, if possible, support the sponsor's formal application for authorization to proceed with the project.

Acquisition Phase
(1) Give Project Manager advice on operational/production matters throughout the phase.
(2) Supply the Project Manager with clauses for inclusion in the Specification(s) relating to operational requirements and others to guide the tenderers from proposing unacceptable designs or procedures.
(3) Carry out preliminary assessments of the tenders from the operational/ production points of view. Advise the Project Manager of discrepancies from the specification(s) and any undesirable features together with the likely costs of correcting these and/or the cost implications of accepting them.
(4) Examine and comment on supplier(s) replies to questionnaires: help the Project Manager to finalize details of contract(s) for the asset.
(5) On the basis of the final contract details, (a) draw up list of production manning requirements and (b) plan a recruitment and training schedule for these personnel, and (c) prepare outline budgets for labour requirements for production and associated (e.g. quality control and safety) staffs.
(6) Put in hand recruitment, training and authorization of staff so that they are ready to undertake operation of asset during plant test and commissioning of the asset.
(7) Check the Technical Manual when received in draft form and produce any further detailed operational procedures necessary.
(8) On receipt of the planned output requirements to meet the expected demands for the product, liaise with the maintenance function to produce complementary production and maintenance plans.

Useful Life Phase
(1) Put asset to work.
(2) Commence operational records for asset and components and monitoring performance.
(3) Review and update plans as necessary.
(4) As required for the safety of maintenance and other personnel, shutdown, isolate and if necessary discharge the asset or components.

Disposal Phase
(1) Shut down, isolate and discharge the asset, systems and components.
(2) Decontaminate throughout where contamination from radioactive or chemically injurious substances exists.
(3) Handover to disposal staff.

Note: For additional aspects of operation/production actions, including communications etc, see BS 3843, Part 3, Section 3, Check List No. 7.

early stages is the valuation of downtime of the asset. In a commercial venture this may be greatly changed if a competitor for the product has been developed by another organization. Two other factors can be involved, both being affected by the overall programme of the associated financial transactions. One involves the timings of actual progress of the raw materials and work in progress and the payments involved. The other is due to the difference in the timing of payments for operating the asset to produce the product and the receipt of income from selling the product. This latter aspect should also take into account the effects of any sales promotions or the granting of credits to purchasers.

Overall, the operational aspects during the useful life phase cannot be considered alone. At all times the mutual effects on or from the concurrent activities of other departments must be considered. Principal amongst these are those of the maintenance department, to be considered in the next chapter. A summary of operational inputs is given in Table 16.1.

Chapter 17

Maintenance Aspects

Objectives

Although maintenance of the asset is a very important aspect of its useful life, it must not be considered as the prime objective during this phase. In fact it represents a cost on the project without directly achieving a return in terms of income or performance. As a result of this the degree of maintenance that should be carried out should be limited to that needed to keep the asset functioning for as long as possible at an acceptable level of performance and at minimum cost consistent with these objectives. Again, as stated earlier, the balance between maintenance cost and performance should be such as to achieve maximum benefit for the project.

It is necessary to make this point as some maintenance specialists have been known to try and extend the maintenance activity as if it were a primary objective in its own right. However, it must be said that in a greater part of industry the opposite is found. Maintenance is neglected both of plant and buildings. For some years in the United Kingdom, the Department of Trade and Industry have been promoting several initiatives to try and improve the level of maintenance. This is primarily to improve the availability of production assets and thereby their productivity and output. Improvements in the economics of the project involved will normally ensue where current levels of maintenance are so far below the levels needed to optimize the project.

Within the remit of achieving the necessary maintenance on the asset and its components, it is still desirable that the maintenance is carried out as economically as possible. In this connection it is seldom that this is dependent upon the direct cost of the maintenance itself. In most cases the

consequential cost of having to shut down the asset to allow this maintenance to take place is more significant. Because of this, carrying out the
maximum amount of maintenance at any single shut down of the asset is
desirable. In addition, the carrying out of maintenance which is possible
and can be done safely with the asset working normally should be exploited
as far as possible. To ensure that the asset and its major components
receive the necessary attention, in the absence of any other need to shut
down production, outages to carry out certain maintenance tasks may have
to be arranged. To get the maximum advantage from such a shut down, as
much maintenance as is possible should be crammed into the period with
priority going to those tasks which can not be carried out with the asset
working. The periodicity of these maintenance shut downs, often termed
overhauls, will be a compromise between the several periods at which
individual maintenance tasks should take place to prevent an unacceptable
loss of performance or an enhanced risk of failure in service.

Development of maintenance practices

In the past, a large proportion of maintenance was of a corrective nature
which would now be classified as 'corrective maintenance'. Even nowadays
we still hear expressions such as 'If it is still working, leave it alone' or 'If it
isn't broken, don't mend it'. This policy of only attending to components
when they break down, invariably leads to a multiplicity of maintenance
outages as individual components fail. Many of these failures and thus the
associated outages could have been prevented by 'catching' them with a
minimum of effort before failure occurred. Furthermore, the extent of
damage and the time taken to restore the asset could be minimized if failure
in service were prevented by repair or replacement of 'worn' components, in
advance of the point in time at which failure would have occurred.

Preventive maintenance became widely advocated in the years following
World War II. The objective was to increase the availability of production
assets by reducing the multiplicity of breakdown outages and minimizing
the downtime for maintenance by periodic inspection of and attention to
components at appropriate intervals of time. For the most part this led to
the majority of these preventive maintenance tasks being established in a
plan, usually a calendar based plan. For many years, where personnel
safety was involved, legislation or statutory regulations had decreed that
certain types of equipment should be examined periodically to detect
degeneration to a dangerous state and should be rectified before being
returned to use. The examination of steam boilers and other specified types

of pressure vessels, are examples of this, as are the more recent examinations of civil aircraft and nuclear reactors.

The periodicity of these statutory examinations often formed the bases of a calendar for the planned maintenance tasks. As many of these tasks as possible are arranged to be carried out concurrently with the major outage for the statutory examination. Those that could not be accommodated within this outage are usually arranged in a small number of outages interspaced between the major statutory examinations. A few tasks can be carried out with the asset running and in production and these will be spaced out throughout the production periods. Should an unforeseen outage for a breakdown occur, the plan should be flexible enough to take advantage of the opportunity to perform some of the preventive tasks concurrently. These will normally be tasks which can be performed within the breakdown outage and are closest to their planned date.

The periodicity of a particular planned maintenance task has to be judged on the basis of the component concerned, its likely failure modes and the expected rate of degenerating mechanisms. To achieve the assurance that the planned maintenance task will 'catch' and prevent failure in service, the periodicity is likely to be set pessimistically at a too frequent level. This can lead to unnecessary effort and costs on the part of the maintenance labour force. As a result, in more recent times, efforts are being made to devise means of detecting whether a task is necessary or not, or at least keeping the work to be carried out to a minimum. Such maintenance is still classified as 'preventive maintenance' but within a sub-classification of 'condition based maintenance'. Typically, the overall task on a component is split into two parts. In the first part, which is incorporated in the calendar plan, the state of the component involved is assessed in some appropriate way. This may be on-load from installed instrumentation or off-load from the results of tests. The second part, attention to the component, is then only performed if the first part identifies it is necessary to keep the item from failing before the next planned date. Such an approach undoubtedly reduces the maintenance effort committed without significantly changing the risks of failure. Most preventive maintenance is being reorganized on the basis of condition based maintenance wherever it is possible to identify clearly the likelihood of failure before the next scheduled date from tests or examinations. This will depend firstly upon there being an appropriate method of testing for this objective and secondly upon the necessary instrumentation being fitted or available. This is an aspect that should be considered carefully by the maintenance function during pre-operational phases, especially the stages of tender assessment and design submission. Later during the 'useful life phase', the possibility of modifying the plant to incorporate devices, from later technologies, to perform the testing task should be considered and incorporated if economically advantageous.

Statutory requirements

Reference has already been made in this chapter to legislation and statutory regulations requiring examination and necessary rectification work on components. In many cases the examinations are detailed and the period between examinations prescribed. More recently the basis of such statutory requirements has been changed to a more flexible one in which the essential principles are laid down which have to be met. This has meant that additional types of plant or equipment are now covered by the regulations. For example in the UK, the Factories Act required such examinations of steam boilers and steam and air receivers. Now the new regulations require all pressure vessels to be examined under the regulations. At the same time the modes of examination have been left to the occupier, usually in collaboration with the inspectorate, to establish the details of the examination methods appropriate to a particular item of plant and the timings of the examinations. Within the limits agreed for the statutory inspections, the user can now maximize the statutory inspection and preventive maintenance tasks into a single outage. An example of how this was achieved under the old type regulations is given by the Certificate of Exception, signed by the Chief Inspector of Factories which allowed the then existing regulations to be modified so that a steam boiler and the steam receivers and air receivers associated with it, could be examined concurrently. Prior to this certificate, the boiler and the receivers examinations were based upon entirely different bases for measuring periodicity and this led to variations in timings for their examinations which therefore could not be kept coincident and several outages were necessary in each examination cycle.

From an economic point of view, failure of a user to comply with the statutory requirements, leaves him open to prosecution and probably large financial penalties. As a consequence, complying with statutory requirements and then optimizing other activities around these, is undoubtedly in the best economic interests of the user's project.

Timing of maintenance

If a user finds that the demand for his product fluctuates in a foreseeable way, such as seasonably, he may find that there are periods when a shut down of the asset has less financial impact on his income than at other times. This should be exploited by arranging his maintenance planning such that the outages for statutory work and/or the necessary outages for planned work take place at such times of low financial impact. Similarly, on a shorter time scale, the impact of an outage at night or at weekends might prove lower than in the day or during weekdays. The possibilities and advantages for maintenance to be carried out at night or at weekends, and the effects of this on maintenance staff working, should be considered when the maintenance manning arrangements are being established prior to or at the beginning of the useful life phase.

Consequential costs of downtime

Although seldom, if ever, allocated against the maintenance function, one of the major costs of carrying out maintenance is the consequential costs of having to shut down the asset to perform certain tasks. It follows that the duration of each maintenance task should be pared down to the minimum. This might be in circumstances in which there would be a slight increase in direct material costs. The reduction in maintenance times may be achieved in many different ways. The first way is to ensure that every task is covered by an appropriate procedural instruction: the most important of these should be given in the Technical Manual supplied by the suppliers. Simpler tasks should still be covered by a written instruction but these can normally be prepared by the user's own maintenance staff. Checking that all these procedures are available should be carried out by the maintenance function prior to start up. The detailed instructions for a task, as given to a maintenance personnel, should expand upon the supplier's manual by identifying the tools and material required and which could be assembled before the asset is shut down. Any preliminary work, which could be done with the asset still running should also be detailed. This would include removal of some protective covers and the erection of scaffolding and platforms. The limits to this on-load work, dictated by safety considerations should also be clearly set down.

Minimizing maintenance downtime

Another ploy to reduce the actual maintenance time can be to carry out repairs by replacements. In this a failed component or, in the case of preventive maintenance, one considered to be at risk of failure, can be removed and replaced by a spare component of the same design or an equivalent functional unit that can be readily fitted. For components which are likely to have a limited life the user will probably carry a number of spares. Replacement by a spare and subsequent fuller tests and inspection of the removed item adds little to the material cost of the maintenance. If the removed item is, in fact, perfectly satisfactory or can be repaired or refurbished to this state it can be returned to the spares pool and re-used at a later date. On the other hand, replacement by a spare, can significantly reduce the outage time. To an extent, this philosophy can be applied to major components but only if an overall economic case can be made for holding a spare in the first place. The economics and practicality of such a policy are discussed in detail in my earlier book, published by Mechanical Engineering Publications, *Strategic Spares and the Economics of Operations*. In the author's experience, outage times have been reduced by months, and in some cases – years, by using this spares exchange policy in very large mechanical and electrical machines. A further advantage of this repair by exchange policy is that the actual repairs or refurbishment processes are carried out with the asset returned to production. The processes can therefore be carried out in a less urgent manner and using the most satisfactory repair techniques rather than the quickest. It can also help to smooth out the work load on the labour resources, leading to possible economies in this area.

Early involvement

Clearly the principal work load on the maintenance workforce occurs throughout the useful life phase, but contributions from the maintenance function are required in the preceding phases. It is therefore highly desirable for an experienced maintenance specialist to be appointed to the project at an early stage, ideally at the same time as the Project Manager. He should then contribute to the specification and subsequently comment, in as much depth as possible, on tenders and design submissions. In doing this, he should have established, at least in outline, the maintenance policy to be adopted for the new asset. This may well be influenced by existing facilities within the user organization or constrained by policy decisions made by the organization's higher management. The essential features which need to be established are the extent of maintenance that can and will be carried out by in-house staff and the maintenance tasks to be carried out by outside agencies. This may include the organization's central workshop, if one exists, the manufacturer or a specialist maintenance contractor. The extent of a repair by exchange policy should normally depend upon the results of economic analyses of spares proposals but experience will normally allow this to be roughly estimated such that a decision of intent can be made on whether or when to adopt such an approach.

Maintainability

During the tender assessment and design submission stages, details of the asset will emerge. The two principal tasks for the maintenance specialist will be to consider the maintainability of the asset and its components and to prepare an initial preventive maintenance plan. As far as maintainability is concerned the principal aspects to be considered are:

(a) Access to the component of the asset concerned – is it feasible? Is it safe?
(b) The tools required to carry out a task – are they standard? If not, are they being supplied?
(c) The materials and spares required to carry out tasks – are these included in a recommended spares list and has this been supplied?

Preventive maintenance plan

A first approach to a preventive maintenance plan can be derived from the application of a procedure known as Reliability Centred Maintenance (RCM). This is already being used by certain organizations and will probably become better known when the international standard guide on this topic, currently being drafted, is eventually published by the International Electrotechnical Commission (IEC). An RCM exercise must, however, be checked and, if necessary, modified to include the examinations required by statutory regulations. The identified tasks, which have to be carried out with the asset shut down, should then be grouped together into a series of maintenance outages. These should be as small a number as possible and of an aggregate length which is a minimum with each outage weighted by their consequential costs of downtime. The supplier should provide the user with a maintenance schedule as part of his Technical Manual. If this has not been based upon his RCM studies, but perhaps by general experience, this should be taken into account in the formulation of the preventive maintenance plan. Later in the pre-production period, the individual tasks in the preventive maintenance plan should have prepared for it a suitable instruction card (or other portable and readily readable device) for the individual maintenance craftsman, on what and how to perform the task, the tools and materials required and the safety measures to be taken. The instruction 'card' could also be devised for the craftsman to report back on the work done, the state of the component and any measurements taken. The Technical Manual, in particular the maintenance instruction section, should be of prime help in this task. As experience builds up in the useful life phase, the preventive maintenance plan and the instructions will need to be reviewed and modified.

Recommended spares

The last stage of the acquisition phase is that of commissioning the asset. Before this takes place, the supplier's Recommended Spares List should have been received and reviewed by the maintenance specialist. He should have placed orders for his stock of routine spares such that they are available from the time of initial start up. Similarly, before start up the user organization should have recruited and, if necessary, trained the key maintenance staff so that these are available from the time the asset is first 'run' in the commissioning stage. Establishing the final numbers of further maintenance personnel required, recruiting them and training them should follow that of the key personnel. Proposals for holding strategic spares should ideally be prepared in this period also.

Corrective maintenance

However well preventive maintenance is carried out, it is unlikely that breakdowns will be entirely eliminated. Indeed if no breakdown occurs on a piece of equipment, it would suggest that the preventive maintenance has been excessive. The maintenance plan should take this into account both in estimating the labour resources needed in total and in the grouping of preventive maintenance tasks and in the allocation of resources when in service.

In the pre-production periods, there is very little than can be done in the way of preparing for these corrective maintenance tasks. However, it should be possible to prepare task instruction cards for elements of the tasks. These sub-tasks would include preparing for access – scaffolding etc., gaining access to a component, removal of a defective item, including necessary dismantling, fitting a replacement or repaired item, re-erection and any testing required. As with the preventive maintenance task instructions, each of these could contain instructions on the tools and materials needed and any safety precautions to be taken. These sub-task cards can then be assembled together with one special card directing what has to be done in the circumstances of that particular failure.

When these cards are used in the useful life phase for both preventive and corrective maintenance, a report back should be encouraged. The report back to be expected from any particular task can usually be anticipated and thus an outline report back can be incorporated in the card which only requires ticks to be made by the craftsman allocated to this task. Space for additional comments should always be provided.

History records

Plant or building history files should be started during commissioning and these returned maintenance reports should be used to keep these histories up to date. When exceptional circumstances have arisen, separate written reports to higher management may be appropriate. In these cases a report back to the appropriate manufacturer/builder or supplier could help in the design of future products. This will normally be based upon agreed arrangements between the user and supplier organizations. Such arrangements help to establish a relationship between the parties which, in the event of problems arising, encourages the supplier to be helpful to the user.

Review of records

At intervals the historical records should be reviewed and analysed from both a qualitative and quantitative point of view. Weaknesses in the design or problems arising from the modes of operation may be detected. Consideration of the latter could support altering certain methods of operating the asset which would ameliorate the problem. Weaknesses in design, on the other hand, should be discussed with the manufacturer who may suggest a modification. Whilst this may appear to resolve the problem, this should not be put in hand automatically. The full performance and economics of effecting the modification should be assessed for viability on a residual life of the asset basis. This assessment should be based upon the optimum time and method for incorporating the modification. The timing should normally be concurrent with an already planned outage of adequate duration. The method may involve, where appropriate and feasible, incorporation in a spare component (off line work) and a subsequent component exchange. These aspects are explained more fully in my earlier book on strategic spares.

Apart from the qualitative analysis of failures from the historical records, a statistical analysis may give a number of quantitative parameters. The numbers of incidents which are available for use in this way, will be severely limited if only a single unit is considered. However, these parameters are well worth calculating for two reasons. The first reason is that the results from the history of the user's own asset(s) can be compared with the published results of other users. Such a comparison will help the user to determine how his asset or component is performing compared with other similar units and whether effort should be made to improve that performance and the likely scope of so doing. Secondly, the user's results can be made available to a party seeking to publish such results and, by adding to the contribution of others, to obtain statistically improved generic data.

The performance parameters to be derived would include 'availability', 'failure rates', and 'mean times to failure and to repair'. It will be noted that some of these parameters, perhaps in qualified terms, will be needed by the user in many of his economic evaluations of alternative courses of action. These include decisions on whether or not to hold certain strategic spares and also on the viability of incorporating modifications.

During the useful life, one of the routine optimization exercises to be performed by the maintenance department is in regard to the stocking and re-ordering of routinely used spares. The factors involved in these exercises include minimum ordering quantities, price discounts for quantity orders, lead times and usage rates. The latter parameter must again be derived from analyses of the historical plant/maintenance records.

Faults

If faults occur in the asset or its components that significantly affect its production performance, these should be investigated to determine the root causes. Reports on the results of these investigations should be included in or attached to the historical record, as well as being distributed to all the interested parties in the user's organization. A copy might be sent to the manufacturer concerned if an appropriate relationship exist between the organizations.

If the fault involves or threatened personnel health or safety, an investigation and report should inevitably result. In these cases, the report may be required by legislation to be sent to the relevant inspectorate or may be called for by the inspectorate at a later date, when they have learnt of the incident. Failure to adequately investigate the incident and its cause could prove very embarrassing to the user.

Tasks performed by operational staff

As mentioned in the last chapter, lubrication of the asset and its components is usually delegated to the operating staff. The schedules of items to be lubricated, the frequency and lubricant to be used should be laid down in written instructions by the maintenance staff. Some form of recording and feedback to confirm that the routines have been carried out is desirable and should be devised. The effectiveness of the lubrication routines should be checked by the examination for wear on the components involved. If problems are found following adoption of the supplier's lubrication recommendations, reference may be made to a tribology specialist to suggest amendments to the routines.

Life limited components

Some components of the asset may have lives less than that of the asset as a whole or of the project it is employed in. Such components will therefore have to be replaced, perhaps more than once, during the asset's life-time. As a modified 'design for life' component cannot be economically or practically devised in all cases, replacements are inevitable. If the component life is predictable, within reasonable limits, dates for replacement exchanges can be set within the calendar of the useful life. These dates should be arranged to coincide with planned outages for other maintenance or statutory work. In such cases, ordering of the replacement can be organized to take place a lead time, plus a small margin, ahead of the replacement date. In other cases the rate of the degeneration mechanism may not be known accurately and the periodic special tests to determine its extent may be necessary. Such tests may themselves be constrained to outages of significant duration which arise only once in several years. The rate of development may therefore be difficult to judge accurately from the few test results available. In such cases the purchase of a strategic spare to be used as a lead replacement unit might be desirable. This approach and the economic justification of it are again discussed in more detail in my earlier book.

Towards the end of the useful life phase, the examinations of the asset and its components take an additional significance. Their state and the likely further working life that they can undertake, should be established. This information should then be used to determine their potential for extending the planned life of the project and the costs of so doing. The economic desirability of extending the project life should be assessed as well as other considerations of optional alternative courses of action.

Obsolescence

Throughout the project the maintenance function should maintain a careful watch on the supply position of essential spares, materials and

Table 17.1 Summary of contributions from maintenance department

Concept Phase

(1) Consider sponsor's perceived outline design for new asset; advise him on any special maintenance aspects and need for maintenance aspects, such as special handling facilities etc. This could include clean conditions workshop or decontamination areas and whether these are already available.

(2) Help to estimate maintenance costs and manpower requirements.

(3) Suggest which strategic spares are likely to be justified and should be included in the initial acquisition contract.

(4) Consider sponsor's formal application for authorization of project and give support if in agreement with it.

Acquisition Phase

(1) Give Project Manager advice on maintenance aspects throughout this phase.

(2) Formulate clauses for inclusion in the specification(s) which:

   (a) identifies the maintenance policy and strategies which will be adopted by the user,

   (b) indicates the user's existing facilities (external workshops etc.) which will be available to the project,

   (c) requires the Technical Manual to cover some or all of the following aspects – maintenance procedures, maintenance schedules, recommended spares lists, erection/installation and handling and storage instructions,

   (d) requires the suppliers to identify all special tools and facilities needed and not already held by the user (see (b)) and to state the period for which he will undertake maintenance on the asset/component supplied,

   (e) requires the supplier's best estimates of the reliability and maintainability of their project and any LCC calculations which are appropriate.

(3) Carry out preliminary assessments of tenders and advise the Project Manager of undesirable features, their cost implications and/or the likely cost of rectifying them; assist in formulating questionnaires.

(4) Examine and comment on replies to questionnaires and with other specialists help the Project Manager to finalize details of the contracts for the asset.

continued

Table 17.1 *Continued*

(5) On the basis of the final contract details (a) plan maintenance personnel recruitment and training ahead of the handover of the asset and components, and (b) put in hand recruitment and training in accordance with the plan.

(6) Plan and commence acquisitions of material, spares and tools etc. required for maintenance.

(7) Check Technical Manual for sections required by maintenance department and prepare detailed procedures where required.

(8) Collaborate with operations/production staff to produce co-ordinated maintenance and production plans.

(9) Arrange training in safety and the user's safety culture for recruited maintenance personnel.

Useful Life Phase

(1) During 'maintenance period' liaise with suppliers with regard to any maintenance required and take over as the period comes to an end.

(2) Commence history records for asset and components from handover or, if possible, from the start of plant testing.

(3) Review and up-date maintenance plans as necessary.

(4) Optimize acquisition plans for spares and materials.

(5) Establish with finance and production departments the 'consequential' costs of 'downtime' of the asset or major components. Use this to review economics of spares plans and proposals.

Disposal Phase

(1) When cleared by production staff, commence dismantling of components earmarked for future use or for special disposal; handover to disposal staff when complete.

Note: For additional aspects of maintenance requirements, including communications etc. see BS 3843, Part 3, Section 3, Check List No. 8.

services, with regard to their continuing availability. Action to counter any premature obsolescence (from the point of view of the user's project) or other causes of inability to supply should start in the acquisition phase. If an inability to supply were to arise in the useful life phase, certain combative actions can be taken as described in Chapter 13 and, in more detail, in my earlier book *Strategic Spares and the Economics of Operations,* 1994, (Mechanical Engineering Publications).

A summary of contributions from the maintenance department over the asset life is given in Table 17.1.

Chapter 18

Disposal Activities

Life expiry

Whilst this chapter, in the main, will deal with the disposal of the asset as a whole at the end of the project life, certain aspects will apply equally to the disposal of components of the asset which are withdrawn (and generally replaced) during the useful life period. There are many reasons why individual components may be withdrawn. These include situations in which the component has reached the end of its individual working life. This may arise because it is patently worn and repair is uneconomic or, in safety situations such as commercial aircraft, it has been assessed or is deemed to be unsafe to give it further operational exposure, lest it fail.

Other situations arise from the displacement of a particular component, and its replacement by one of up-dated design as a result of an authorized modification. Finally, a component may have to be replaced as a result of a failure leading to either direct or consequential damage to the component.

Possible reuse of components

When a complete asset or a component is withdrawn at the end of its project life for which it was purchased, the possibility of it being used further, or parts of it being used further should be established. This will include being used in any context, not necessarily the same as that for which it was used in the original project. If sold, the income from the sale should be credited to the original project. On the other hand, if it is to be used in a further project of the user organization, a 'book' transfer at an agreed value should take place with the value credited to the original project and debited against the new project. The 'agreed' value will normally be one agreed between the two Project Managers concerned or dictated by their higher management.

Economic management of disposals

The principal aim of the Project Manager will be to obtain the maximum credit from the sale of the disposed item or items. This may be obtainable through splitting an item into several portions rather than as a whole. However, the cost of dividing the item and disposing of the parts separately must also be taken into account and the net credit obtainable should be the determining factor on the methods to be adopted.

If an item or items can be neither sold nor re-used by the organization, its disposal should still be made in as profitable way as possible. One way of obtaining some value from it is by selling it as scrap materials. Modern technology has developed by using relatively new materials in some components. Such materials tend to be in limited supply and can be expensive. Thus recovery of these materials can be a profitable business on its own albeit, in many cases, specialized. However, there can be significant rewards from handling such materials in a way that allows them to be extracted by the purchaser of the scrap. To allow this, the existence of rare or specialized materials in the design of components must be confirmed and this should be obtainable from the original or revised technical manual for the asset. Then, as dismantling of the asset and components is undertaken, the materials should be appropriately segregated to allow the different forms of scrap to be handled and sold separately to appropriate buyers. The effort and cost of doing this must, of course, be reviewed in relation to the net income or cost involved.

As has been noted in this book, economic analyses should be carried out of the complete project and, more particularly, the expected costs and income for the remaining planned life of the project, As the project approaches the end of the originally planned life, these later assessments need to be reviewed in the light of the two alternatives of extending the planned project life or of prematurely closing down the project and, possibly replacing it with a new project. These assessments should ideally be carried out before committing the project to some major expenditure such as the replacement of a key expensive component or a major overhaul. These assessments should take into consideration the likely costs, and returns of the disposal activities. This arises because, being much closer in time, these will have a proportionately greater effect on the economics. A principal factor in considering these variations from the original planned life of the project will be the continuing need for the project's product and its marketability. When considering extending the project life a major factor will be the ability of the asset and its components to survive the extended period. Whilst many items may easily survive the further period, others may become life expired and thereby incur exceptional costs for their replacement. In turn these may have a significant effect on the disposal plans and costs. Another aspect of extending the life and the actions necessary on the life expired components will be the ability to replace them. Often, in the period since their first acquisition, a number of these components will have been declared obsolete. The possibility of dealing with such situations has been discussed earlier, but again the costs of doing so may, at this late stage in the originally planned project life, be economically prohibitive and the concept of extending the life rendered impracticable.

Safety

Throughout the whole of the disposal activities the principal concern must be one of safety, both for those taking part in the activities and for the general public.

Let us first consider the initial stage of the disposal activity, namely the dismantling of the asset. For various reasons this activity is frequently subcontracted to an external firm rather than left to the organization's own personnel. If it is left to the organization's own personnel, these will normally be drawn from their maintenance staff who, in the case of plant and equipment, may well have experience in dismantling these items for maintenance during the useful life. However, if it is delegated to a sub-contractor, this experience will be lacking and thus extra precautions should be taken for their safety. Dismantling instructions should have been included in the asset's Technical Manual and the user should ensure that copies of relevant sections are passed to the sub-contractor. These should include references to safe methods and hazards, such as exceptionally heavy weights requiring a crane or lifting devices. Another potential source of hazard, which may not be explained in the Technical Manual, will be chemical or radioactive contamination arising from the use of the asset. Clearly either this will require that the items be decontaminated before handing over for dismantling or the sub-contractor should have these hazards drawn to his attention and it should be established that he has the necessary knowledge and experience to deal with them.

Hazardous materials

Hazardous materials used in the manufacture of the asset or components themselves should have been identified in the Technical Manual. However, care should be taken in case such materials have been introduced as a result of a modification, and not recorded, or, as a result of experience, an original material has been reclassified dangerous since first being incorporated. Asbestos, originally used extensively as a thermal insulant, is a good example of such a reclassification.

Buildings and civil structures

With buildings and civil structures the situation is rather different. Firstly, many buildings are capable of being adopted to other uses and thereby become capable of being re-used by the user organization or sold to another. Specialist structures such as a chimney, together with any unwanted buildings, will generally need to be demolished so as to give a clear site for any future owner or user of the land it occupied. Such work tends to be rather specialized and needs specialist equipment. As a result, it is normally left to a specialist sub-contractor who will be familiar with the hazards and the safety precautions required.

Disposal criteria

Finally, items and materials which are no longer required by the user organization and which can not be sold either for re-use or as scrap, will have to be disposed of. This should be carried out in a way that best accords with two criteria. Firstly, they must be disposed of in a way that does not create any environmental hazard. Secondly, within the acceptable methods of disposal the method used should be the one which has the lower costs. Although some materials can not be sold as scrap, if they are recyclable, it may be possible to offer these to an organization which would arrange for their collection at no or very low costs. Material now deemed to be hazardous may have to be sent to specialized disposal sites and treated or packaged in a specially ordered manner before despatch. Inevitably this will involve some costs but the costs of failing to obey associated regulations will be much greater.

Appendix I

Glossary of terms

The majority of technical terms used in this book are based upon the definitions given in British Standard BS 3811: 1993, *Glossary of terms used in terotechnology* (see Appendix II). In the economic management of assets certain terms are used in specific ways. The following glossary is presented for the convenience of readers and shows these specific uses.

Term	*Definition*
Acquisition phase	The period of the project life in which actions take place directly concerned with the supply of the asset. It commences with the preparation of the specification and continues through the design, development, manufacture, erection and commissioning stages up to the handover to the user.
Asset	The buildings, plant, machinery and all other permanent items required by the user to produce and supply the product.
Availability	The ability of an item to perform its required function over a stated period of time (weighted by the rate of actual output compared with the asset's rated output).
Concept phase	The initial period in the project life in which the outline of the project and its product is devised and its economics estimated and presented for formal approval to proceed.

Downtime	The period of time during which an asset is not in a condition to perform its intended function.
Failure	The termination of the ability of an item to perform a required function.
Feed	The material etc. from which the user produces the product within or by use of the asset.
Lead time	The time required to acquire a given item or stock, measured from the time that the acquisition is approved until it is delivered.
Life cycle	The time interval that commences with the initiation of the concept for the project and terminates with the disposal of the asset.
LCC (life cycle cost)	The aggregated costs of the project over its total life.
Product	A generalized term for the output from the asset.
Project	A generalized term for the proposal to provide and supply a product and acquire the means to do so.
Service	An essential input to the asset which is neither incorporated in the asset nor in the product.
Spare	A component or sub-assembly held available for maintenance purposes or for the replacement of faulty parts.
Terotechnology (economic management of physical assets)	A combination of management, financial, engineering, building and other practices applied to physical assets in pursuit of economic life cycle costs.
Useful life phase	The period of the project life in which the user operates the commissioned asset to produce the product.

Appendix II

Bibliography

A number of references are made in this book to an earlier work of the author. This earlier book deals in some depth with a particular application of the philosophy of terotechnology with which this book is concerned. The earlier book is:

HODGES, N.W., 1994, *Strategic spares and the economics of operations*, Mechanical Engineering Publications Limited, London.

References are also made to a number of publications by the British Standards Institution. These are:

BS 3811: 1993. *Glossary of terms used in terotechnology*

BS 5760. *Reliability of systems, equipment and components*. This is a multi-part standard comprising parts issued at different dates, each part dealing with a separate aspect of reliability work such as Part 7, 'Guide to fault tree analysis'. Several parts are identical to or are derived from international standards issued by the IEC.

BS 5750: *Quality systems*. This is again a multi-part standard dealing with the various aspects of quality systems. Certain parts are identical with the international standards in the ISO 9000 series and the European EN 29000 series.

Although not referred to in this book, a number of other British Standards deal with the practice of economic management of assets or the techniques which are incorporated in its application. These include:

BS 3843: 1992/3. *Guide to terotechnology (the economic management of assets)*. This is in three parts – Part 1: 1992 'Introduction to terotechnology', Part 2: 1992 'Introduction to the techniques and applications', Part 3: 1992 'Guide to the available techniques'.

Index